51-74 DEV
MATH

```
*****        RAL LI
Acc_No:    15004
Shelf: 51-74 DEV
       E61
 ISBN: 0045150036
Copy:    1
```

LIBRARY
RUTHERFORD
18 MAY 1987
LABORATORY

Numerical Methods in Engineering & Science

Titles of related interest

Analytical and computational methods in engineering rock mechanics
E. T. Brown (ed.)

Boundary element method in solid mechanics
S. L. Crouch & A. M. Starfield

The boundary integral equation method for porous media flow
J. A. Liggett & P. L-F. Liu

Computers in construction planning and control
M. J. Jackson

The finite element method in thermomechanics
T-R. Hsu

Numerical Methods in Engineering & Science

Graham de Vahl Davis
*School of Mechanical and Industrial Engineering,
University of New South Wales,
Kensington, NSW, Australia 2033*

London
ALLEN & UNWIN
Boston   Sydney

© G. de Vahl Davis, 1986
This book is copyright under the Berne Convention.
No reproduction without permission. All rights reserved.

Allen & Unwin (Publishers) Ltd,
40 Museum Street, London WC1A 1LU, UK

Allen & Unwin (Publishers) Ltd,
Park Lane, Hemel Hempstead, Herts HP2 4TE, UK

Allen & Unwin, Inc.,
8 Winchester Place, Winchester, Mass. 01890, USA

Allen & Unwin (Australia) Ltd,
8 Napier Street, North Sydney, NSW 2060, Australia

First published in 1986

British Library Cataloguing in Publication Data

De Vahl Davis, Graham
   Numerical methods in engineering and science.
1. Mathematics   2. Numerical calculations
I. Title
511′.0245   QA297
ISBN 0-04-515002-8
ISBN 0-04-515003-6

Library of Congress Cataloging in Publication Data

Davis, Graham de Vahl.
   Numerical methods in engineering and science.
Bibliography: p.
Includes index.
1. Engineering mathematics.   2. Science—Mathematics.
3. Numerical analysis.   I. Title.
TA335.D38   1986   620′.0042   86-17487
ISBN 0-04-515002-8 (alk. paper)
ISBN 0-04-515003-6 (pbk. : alk. paper)

Set in 10 on 12 point Times by Paston Press, Norwich
and printed in Great Britain by Anchor Brendon Ltd,
Tiptree, Essex

To the memory of my parents:
Rose and Gerald de Vahl Davis

and to my wife and daughters:
Vivianne, Shelley and Nicola

Preface

This book is designed for an introductory course in numerical methods for students of engineering and science at universities and colleges of advanced education. It is an outgrowth of a course of lectures and tutorials (problem-solving sessions) which the author has given for a number of years at the University of New South Wales and elsewhere. The course is normally taught at the rate of $1\frac{1}{2}$ hours per week throughout an academic year (28 weeks). It has occasionally been given at double this rate over half the year, but it was found that students had insufficient time to absorb the material and experiment with the methods. The material presented here is rather more than has been taught in any one year, although all of it has been taught at some time.

The book is concerned with the application of numerical methods to the solution of equations – algebraic, transcendental and differential – which will be encountered by students during their training and their careers. The theoretical foundation for the methods is not rigorously covered. Engineers and applied scientists (but not, of course, mathematicians) are more concerned with using methods than with proving that they can be used. However, they must be satisfied that the methods are fit to be used, and it is hoped that students will perform sufficient numerical experiments to convince themselves of this without the need for more than the minimum of theory which is presented here.

The emphasis, as far as differential equations are concerned, is towards finite difference methods, which form the basis of most introductory courses on numerical techniques. Weighted residual, finite element and boundary solution methods are briefly introduced, as students should at least be aware of these important procedures. I would like to thank my colleague Dr Don Kelly for contributing the major part of the chapter on these integral methods for boundary value problems. The depth of coverage given to partial differential equations (especially hyperbolic equations) is rather less than that given to ordinary differential equations. However, the material included would be suitable for leading to more-advanced courses, such as one in computational fluid dynamics. The choice of topics is somewhat subjective, but it is hoped that those selected for inclusion cover the basic material needed in a first course.

A number of worked examples and problems is given. It cannot be emphasized too strongly that students will understand numerical methods only once they have used them. Therefore, those who teach numerical methods are urged to seek the co-operation of their colleagues in other subjects in the construction and setting of problems which demand the use of these methods. Students sometimes have difficulty in synthesis: i.e. in the bringing together of the various strands of their course. Lecturers can help by actively seeking 'interdisciplinary' exercises. Numerical methods are just a tool, not an end in themselves, and are intended to be used in conjunction with analytical aspects of solid mechanics, fluid mechanics, heat transfer, etc.

Students are urged to seek out or construct further problems for themselves. It is often easy to design a problem with a known answer, and even if a problem without a known answer (such as a difficult differential equation) is attempted, substitution of the answer back into the question will normally verify (or falsify!) the answer.

Many of the methods can be successfully illustrated and used with just a simple electronic calculator; other methods involve so much computation that a programmable device is needed; and still other methods will require a digital computer. It is hoped that students will have (almost) unlimited access to a digital computer – enough so that they will be able to experiment freely with the various techniques, but sufficiently restricted that they will accept that computing power is not an infinite resource but one which, in the real world, must be paid for and therefore used carefully.

In the examples illustrated with computer programs, Fortran has been used. The programs may need slight modification to take into account local dialects and operating systems, and file definition statements for 'hard copy' input and output have been omitted since these are system-dependent. It is believed that the programs are sufficiently readable that they can be implemented without much difficulty on any machine.

Graham de Vahl Davis

Contents

Preface	*page*	ix
List of tables		xv

1	**Introduction**		1
	1.1	What are numerical methods?	1
	1.2	Numerical methods versus numerical analysis	3
	1.3	Why use numerical methods?	4
	1.4	Approximate equations and approximate solutions	5
	1.5	The use of numerical methods	6
	1.6	Errors	8
	1.7	Non-dimensional equations	11
	8	The use of computers	12

2	**The solution of equations**		14
	2.1	Introduction	14
	2.2	Location of initial estimates	15
	2.3	Interval halving	19
	2.4	Simple iteration	24
	2.5	Convergence	26
	2.6	Aitken's extrapolation	32
	2.7	Damped simple iteration	34
	2.8	Newton–Raphson method	37
	2.9	Extended Newton's method	43
	2.10	Other iterative methods	45
	2.11	Polynomial equations	47
	2.12	Bairstow's method	56
		Worked examples	58
		Problems	64

3	**Simultaneous equations**		71
	3.1	Introduction	71
	3.2	Elimination methods	73
	3.3	Gaussian elimination	75
	3.4	Extensions to the basic algorithm	80
	3.5	Operation count for the basic algorithm	81

3.6	Tridiagonal systems	83
3.7	Extensions to the Thomas algorithm	86
3.8	Iterative methods for linear systems	89
3.9	Matrix inversion	94
3.10	The method of least squares	96
3.11	The method of differential correction	100
3.12	Simple iteration for non-linear systems	103
3.13	Newton's method for non-linear systems	106
	Worked examples	108
	Problems	113

4 Interpolation, differentiation and integration — 116

4.1	Introduction	116
4.2	Finite difference operators	118
4.3	Difference tables	123
4.4	Interpolation	125
4.5	Newton's forward formula	126
4.6	Newton's backward formula	130
4.7	Stirling's central difference formula	131
4.8	Numerical differentiation	132
4.9	Truncation errors	134
4.10	Summary of differentiation formulae	136
4.11	Differentiation at non-tabular points: maxima and minima	138
4.12	Numerical integration	139
4.13	Error estimation	142
4.14	Integration using backward differences	142
4.15	Summary of integration formulae	143
4.16	Reducing the truncation error	146
	Worked examples	149
	Problems	153

5 Ordinary differential equations — 157

5.1	Introduction	157
5.2	Euler's method	158
5.3	Solution using Taylor's series	163
5.4	The modified Euler method	165
5.5	Predictor–corrector methods	168
5.6	Milne's method, Adams' method, and Hamming's method	170
5.7	Starting procedure for predictor–corrector methods	172
5.8	Estimation of error of predictor–corrector methods	174
5.9	Runge–Kutta methods	176
5.10	Runge–Kutta–Merson method	179
5.11	Application to higher-order equations and to systems	180
5.12	Two-point boundary value problems	186
5.13	Non-linear two-point boundary value problems	198
	Worked examples	199
	Problems	205

6 Partial differential equations I – elliptic equations — 210

- 6.1 Introduction — 210
- 6.2 The approximation of elliptic equations — 212
- 6.3 Boundary conditions — 214
- 6.4 Non-dimensional equations again — 215
- 6.5 Method of solution — 217
- 6.6 The accuracy of the solution — 221
- 6.7 Use of Richardson's extrapolation — 222
- 6.8 Other boundary conditions — 223
- 6.9 Relaxation by hand-calculation — 225
- 6.10 Non-rectangular solution regions — 231
- 6.11 Higher-order equations — 238
- Problems — 239

7 Partial differential equations II – parabolic equations — 243

- 7.1 Introduction — 243
- 7.2 The conduction equation — 243
- 7.3 Non-dimensional equations yet again — 244
- 7.4 Notation — 245
- 7.5 An explicit method — 246
- 7.6 Consistency — 251
- 7.7 The Dufort–Frankel method — 252
- 7.8 Convergence — 253
- 7.9 Stability — 256
- 7.10 An unstable finite difference approximation — 260
- 7.11 Richardson's extrapolation — 261
- Worked examples — 262
- Problems — 265

8 Integral methods for the solution of boundary value problems — 267

- 8.1 Introduction — 267
- 8.2 Integral methods — 267
- 8.3 Implementation of integral methods — 271
- Worked examples — 278
- Problems — 281

Suggestions for further reading — 283

Index — 284

List of tables

1.1	Round-off errors	page 9
1.2	Reduction in truncation error in the evaluation of the sum of the series (1.8)	10
1.3	The truncation error in an iterative process	11
2.1	Values of $f(x) = x^2 - \sin x - 5$	16
2.2	The use of (2.24) to estimate the error in successive estimates of the solution of $x^3 - 3x^2 - 3.88x + 3.192 = 0$	30
2.3	The use of (2.24) to estimate the error in successive estimates of the solution of $x^3 - 3x^2 - 3.88x + 12.824 = 0$	30
2.4	Effect of λ on the rate of convergence of (2.29) when used to solve (2.14)	36
2.5	Comparison of the solution of $x^2 - 3x + 2 = 0$ by simple iteration and by Newton's method	43
3.1	Solution of (3.28) by Jacobi iteration	91
3.2	Solution of (3.28) by Gauss–Seidel iteration	92
3.3	Use of under-relaxation to achieve convergence	94
3.4	Least squares analysis of data	99
3.5	Experimental data to be represented by $y = ax^b$	100
3.6	Curve-fitting by the method of differential correction	103
3.7	Calculations for Worked Example 3	110
4.1	Hypothetical experimental data	116
4.2	The relationships between the finite difference operators	122
4.3	A forward difference table based on Table 4.1	123
4.4	A backward difference table based on Table 4.1	124
4.5	A difficult interpolation problem	128
4.6	Three-point formulae ($n = 2$)	137
4.7	Four-point formulae ($n = 3$)	137
4.8	Five-point formulae ($n = 4$)	138
4.9	Data for the Worked Examples	149
4.10	The forward difference table from the data of Table 4.9	149
4.11	Data to test Richardson's extrapolation for numerical differentiation	151
5.1	A comparison of the errors in some methods for the solution of (5.11)	166
5.2	The use of Richardson's extrapolation to improve the accuracy of the finite difference solution of a differential equation	192

5.3	An illustration of the shooting method for $y'' + 2xy' - 6y - 2x = 0$ with $y(0) = 0$ and $y(1) = 2$	197
5.4	Modified Euler solution for Worked Example 2	201
5.5	Fine mesh starting values for Worked Example 3	202
5.6	Continuation of the solution of Worked Example 3	202
7.1	The explicit solution with $r = 0.5$	249
7.2	The explicit solution with the mesh sizes halved	250
7.3	Error growth using the explicit method with $r = \frac{1}{2}$	256
7.4	Error growth using the explicit method with $r = 1$	257

1

Introduction

1.1 What are numerical methods?

It is not easy to explain what is meant by the term 'numerical methods' except by giving examples – and the rest of this book is concerned with little else but examples. Therefore the simplest, but tautological and rather unhelpful, answer to this question is 'they are what this book is about'. More precisely, but perhaps not much more helpfully, they are methods which can be used to obtain numerical answers to problems when, for one reason or another, we cannot or do not wish to use analytical methods.

A simple example is provided by the quadratic equation

$$ax^2 + bx + c = 0 \tag{1.1}$$

The *analytical* solution to this equation is

$$x = \{-b \pm \sqrt{(b^2 - 4ac)}\}/2a \tag{1.2}$$

There are, in fact, two solutions for any given set of values of a, b and c, and (1.2) may be used to evaluate these solutions. This process of evaluation, although involving 'numbers' rather than symbols, is *not* a numerical method.

On the other hand, suppose we do not know how to evaluate a square root. Then (1.2), although a formally correct statement of the solution of a quadratic equation, will not be of any use to us and we will have to find another way to solve (1.1). One way of doing this is to rewrite it in the form

$$x = -(bx + c)/ax \tag{1.3}$$

Equation (1.3) is not an analytical solution in the way that (1.2) is. The right-hand side still involves x, the unknown, and therefore cannot be evaluated. However, (1.3) can sometimes be used to find x for particular values of a, b and c.

Consider, for example, what happens when $a = 3$, $b = -5$ and $c = 2$. Then (1.3) becomes

$$x = (5x - 2)/3x \tag{1.4}$$

We now guess a value for x, substitute this into the right-hand side of (1.4) and see if the value we calculate is equal to the one with which we started. (We will see later how to do better than merely guessing.) It is most unlikely that it will be, but under the right circumstances (discussed later in the book) the calculated value will be a *better estimate* of the value of x.

Suppose we guess that $x = 2$. When this value is put into the right-hand side of (1.4), we find that the new value of x is 4/3. If this were really the solution of the equation, then substituting it into the right-hand side of (1.4) should again yield 4/3. However, on doing this we find that we get 7/6. Perhaps *this* is the answer. But no – on insertion of this value into the right-hand side of (1.4) we do not get 7/6 again, but 1.0952 (approximately). Continuing this process of calculation and resubstitution, we find that we always get slightly different answers, the next values being 1.0580, 1.0365, 1.0235, etc. If we had the patience and the time, we would find that our 'answers' were getting closer and closer to what (1.2) tells us the *correct* answer is, namely 1. We would also see that, providing we work with sufficient numbers of decimal places*, we would never actually get there.

This is a 'numerical method' for solving (1.1). It exemplifies many of the characteristics of the numerical methods that we will be discussing in later chapters.

First, we can only use this method if actual numerical values are known for a, b and c: otherwise the right-hand side of (1.3) cannot be computed.

Secondly, it is an *iterative* method: it involves the repetitive application of the same arithmetic operations with different data, generally for an unpredictable number of times.

Thirdly, it does not give us the exact solution – it only yields estimates which get successively closer to the exact solution.

Fourthly, it does not always work! (Try it with $a = 1$, $b = -3$ and $c = 2$. Use $x = 0.99$ – which is almost the correct answer – as the first guess for x. The condition under which we can be certain that this particular method will work is discussed in Section 2.5.)

Not all numerical methods are iterative. Some can be classed as *direct* methods. In these, a prescribed sequence of arithmetic operations is executed which is not (or at least, not entirely) repetitive. For example, consider a *system* of equations such as

$$x_1 + 5x_2 = 7 \qquad (1.5a)$$

$$3x_1 - x_2 = 5 \qquad (1.5b)$$

---

* Of course, there is always a practical limit to the number of decimal places that we can retain, so eventually the answers *will* stop changing within the accuracy available.

This system can be solved by combining the equations in such a way as to eliminate one of the two unknowns. If the second equation is multiplied by 5 and the two equations then added, x_2 will be eliminated and the resulting equation can be solved for x_1. The other unknown, x_2, can then be found by inserting the now known value for x_1 into either of the original equations. This process is a *direct* method for solving a system of equations. We will see later that there are also *iterative* methods for systems.

1.2 Numerical methods versus numerical analysis

This book is intended for students meeting numerical methods for the first time in their undergraduate course. It therefore does not place a great deal of emphasis on the *analysis* of those methods – that is, on the proof that the methods work and are fit to be used. Instead, we shall usually ask the student to take them on trust – to accept the evidence of the examples presented, and of the examples the students work for themselves, Numerical analysis is a proper and important part of mathematics. However, for undergraduates in the early part of their course, and especially for undergraduates who are not specializing in mathematics, it can be an abstract and difficult subject which tends to discourage the prospective practitioner of numerical methods. (It is also true that the analysis of numerical methods tends to lag somewhat behind their development, so that they are sometimes used before a full proof of their validity has been found. However, this is not the case with the methods presented in this book.)

Most of the *methods* presented here will not therefore be accompanied by *analysis*. Where restrictions apply we shall discuss them and, when it is felt to be useful to the discussion, we shall *prove* that these restrictions exist: i.e. we shall perform some simple numerical analysis. However, the emphasis here is on the *use* of methods, and their detailed analysis is left for further studies by those students who are so inclined.

In a similar vein, many questions which concern mathematicians, such as the existence and uniqueness of solutions, are side-stepped completely. This is a book for engineers and applied scientists, and a pragmatic viewpoint has been adopted: it is assumed that problems in engineering and applied science *do* have solutions, and that they generally have *unique* solutions. The former assumption implies that the engineering problem has been correctly expressed as a mathematical problem – an aspect to which we shall give some attention, although it is really beyond the scope of this book, and is more a matter for consideration in the subjects to which these methods are to be applied. The latter assumption – that the solution is unique – is not always true*, but it is *generally* true and we will therefore not cloud our discussion of solution methods with questions of existence and uniqueness.

1.3 Why use numerical methods?

In general, analytical methods for the solution of the equations which arise in applied science or engineering are to be preferred, as they lead to *general* rather than *particular* solutions. Thus, (1.2) is a general solution to (1.1), and can be used for any set of values of a, b and c. On the other hand, the numerical method illustrated in Section 1.1 must be repeated in its entirety whenever different values of a, b or c are required. Moreover, analytical solutions give more information: the nature of the dependence of x on a, b and c is revealed by (1.2), whereas this dependence could only be discovered from a numerical solution by the tedious process of using a range of values of the parameters, together with graphical or other means to discover a pattern in the solutions.

However, there are some situations in which a numerical method may be preferred despite an analytical solution being available. These arise when the analytical solution is such that its evaluation is extremely time-consuming. For example, if the solution to a problem involves a complicated series which converges very slowly, and which cannot be summed analytically, the practical evaluation of the solution may be more efficient by a numerical method than by the computation and summation of many terms in the series. Fortunately, these situations are relatively rare, and it is again stressed that analytical solutions are normally preferable to those obtained numerically.

Unfortunately, engineering problems are generally highly complex, often involving non-linear phenomena, and it is not uncommon to find that our mathematical knowledge is not sufficient to enable an analytical solution to a 'real' problem to be found. We only have to make a small alteration of (1.1) to

$$ax^{2.1} + bx + c = 0$$

---

* For example, the nature of the flow of a fluid in a pipe – whether the flow is smooth and steady (*laminar* flow) or randomly fluctuating (*turbulent* flow) – depends on a quantity known as the Reynolds Number (see Section 1.5). For low values of the Reynolds Number the flow is laminar, and for high values it is turbulent. However, if the Reynolds Number is at the transition value (which is approximately 2100) the flow may be either laminar or turbulent, depending on upstream conditions and other factors, and there are therefore two possible solutions of the equations of motion. Similarly, a strut or column which deflects under a given compressive load may assume any one of several shapes. Such situations are often associated with a stability problem. Thus, while it is possible experimentally to achieve laminar flow in a pipe at a Reynolds Number greater than 2100, such a flow is marginally unstable and the introduction of a small disturbance may cause transition to turbulence. Nevertheless, the laminar and turbulent velocity distributions are *each* correct solutions of the equations of motion. The student should be aware that there may be situations in which the solution obtained is not the only one, nor the one which would actually be observed in nature.

to obtain an equation that cannot be solved analytically. When we come to the solution of the differential equations which describe the real world, and which are often non-linear, it is almost the exception to find a problem which *can* be solved analytically without making some simplifying assumptions.

Another situation in which numerical methods must be used is when the information being handled is given in tabular, rather than in functional, form – for instance, when it has been obtained in an experiment. Any treatment of this information (its differentiation or integration, for example, or perhaps its use as a forcing function in a differential equation) must be by a numerical method.

Although many of the examples used in this book are capable of being solved analytically, students should realize that they have been chosen only so that a comparison may be made between the analytical and the numerical solutions, in order to provide some idea of the reliability of the latter.

1.4 Approximate equations and approximate solutions

When, as is often the case with a problem arising from nature, our mathematical knowledge is inadequate to allow solution of the appropriate equations which describe the problem, there are two approaches which may be taken. The first is to seek to change the equations into a more manageable form. This is done by the elimination of certain terms which are believed to have only a small effect on the solution compared with the effect of other terms, or by linearization of some of the non-linear terms, or by other means. The result is an equation or system of equations which *can* be solved analytically, and it might be said that this process leads to an *exact* solution of an *approximate* equation. However, this solution is suspect, since one often cannot be certain of the effect of the simplifying assumptions without making a comparison with experimental data. The accumulation of experience of the results of such simplifications and comparisons strengthens confidence in this approach, but caution usually dictates that some experimental evidence must be provided in support of any new analytical solution.

On the other hand, using methods such as those described in this book, 'solutions' can *generally* be obtained to the full equation or equations without the need for any simplifying assumptions. Such solutions are only approximations – the method outlined at the beginning of this chapter leads only to an approximation to the solution of (1.1), although the accuracy of the approximation may be improved as much as we wish simply by doing further calculations. Thus, it might be said that numerical methods lead to an *approximate* solution of an *exact* equation. This solution is not merely suspect – we *know* that it is wrong! However, and this is the key point, we can often make an assessment of the magnitude of the error without recourse to experiment, and we can always improve the accuracy by investing more effort (which means, in most cases, more money for computer time).

It should not be inferred from these remarks that numerical methods have supplanted analysis. It has already been stressed that analytical solutions possess greater generality and disclose more information than numerical solutions do. However, it is true that the latter are usually more trustworthy, and the need for their experimental verification is continually diminishing. Indeed, it might be mentioned here that numerical 'experiments' are often preferable to physical experiments. They can be cheaper, they can be performed more rapidly, any of the variables can be 'measured' as precisely as the approximations of the method (i.e. as the budget) permits, and all parameters thought to be significant can be accurately controlled. There are some who would argue with this proposition, but there are also many who believe it to be true. The numerical study of unexplored physical problems is assuming a growing importance in the field of scientific activity.

1.5 The use of numerical methods

There are three stages in the development of a numerical method for the solution of an engineering or scientific problem:

- the formulation of the problem
- the development of a mathematical model
- the construction of an algorithm.

Suppose it is desired to study the motion of a body falling vertically under the influence of gravity in a viscous liquid. In *formulating the problem* the factors influencing the motion must be recognized: i.e. the weight of the body, and the buoyancy and drag forces exerted by the fluid on the body. The density and viscosity of the fluid will be relevant, as will the shape of the body. These factors will be interrelated, and it will be necessary to have an understanding of the fluid mechanics of the problem in order to determine these relationships.

The *development of a mathematical model* starts with the expression of these relationships in symbolic form. Let the net force acting vertically downwards on the body be F, the weight of the body be W, the buoyancy force be B and the drag force be D. These are the only forces acting, so

$$F = W - B - D$$

The weight of the body is its mass m multiplied by the local gravitational acceleration g. The buoyancy force, in terms of m, ρ and ρ_1, the densities of the body and the liquid, respectively, is $m(\rho_1/\rho)$. The drag force is usually expressed in terms of a drag coefficient C_D, the velocity of the body v and its cross-sectional area normal to the flow direction A, as $C_D(\frac{1}{2}\rho_1 v^2)A$. Then

$$F = mg - m(\rho_1/\rho)g - C_D(\tfrac{1}{2}\rho_1 v^2)A$$

Since the net force F is also equal to the acceleration dv/dt of the body times m, we obtain

$$\frac{dv}{dt} = \left(1 - \frac{\rho_1}{\rho}\right)g - \frac{C_D \rho_1 A}{2m} v^2$$

To make further progress, we need to know something about the drag coefficient. This turns out to be a complicated function of the size and velocity of the body, and of the density and viscosity of the liquid. These quantities can be combined into a dimensionless number called the Reynolds Number and denoted by Re:

$$\text{Re} = \rho_1 v L/\mu$$

where L is a *characteristic dimension* of the body (i.e. a dimension which is relevant to the problem under consideration) and μ is the viscosity of the liquid. The nature of the functional relationship between C_D and Re depends on the shape of the body and its orientation to the direction of motion, and cannot generally be expressed in a simple analytical form. For example, Figure 1.1 shows the relationship for a sphere of diameter L. Except for small values of Re, a theoretical expression for C_D cannot be found, and Figure 1.1 is the result of experiments.

By methods such as those to be described in Chapter 3, an empirical relationship

$$C_D = f(\text{Re})$$

could be found for a body of a given shape, enabling the equation for the motion of the body to be written:

$$\frac{dv}{dt} = \left(1 - \frac{\rho_1}{\rho}\right)g - \frac{\rho_1 A}{2m} f(\text{Re}) v^2 \tag{1.6}$$

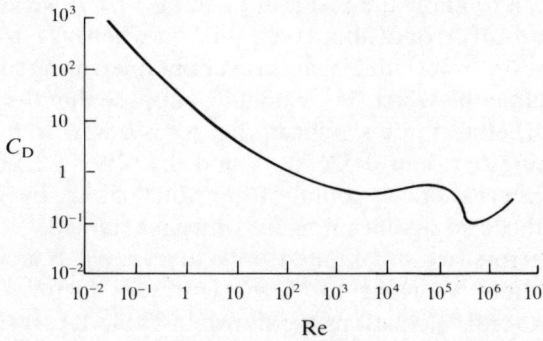

Figure 1.1 The drag coefficient of a sphere versus Reynold's Number.

Equation (1.6) is a first order ordinary differential equation. To solve it, an initial condition is required. If the falling body was released from rest, then the mathematical model is completed by the requirement

$$v = 0 \quad \text{at } t = 0 \tag{1.7}$$

The third stage in the development of a numerical method is the prime concern of this book – it is the *construction of an algorithm*, i.e. of the series of step-by-step operations which must be performed in the actual implementation of a particular numerical method. The need for a numerical method must first be verified: it must be ensured, for example, that $f(\text{Re})$ is such that an analytical solution of (1.6) cannot be found. Having established the need for a numerical solution, we may consider which of several alternative methods would be the most appropriate. Methods for the numerical solution of this differential equation are described in Chapter 5.

1.6 Errors

Almost all calculations involve errors. If we wish to add 1 to 7 we can obtain the exact answer (8), but if we wish to divide 1 by 7 the exact answer is a decimal fraction with an infinite number of figures. We have to be content with an approximate answer – say 0.143 – which is in error by about 0.0001428.... We have incurred this error because, in practice, we can only cope with a finite number of significant figures: in this case three. All calculating devices, from pencil-and-paper to the most sophisticated computer, have this restriction. Typically, hand-held calculators are limited to eight or ten significant figures; large, high-speed computers can handle a few more.

If the result of a calculation requires more significant figures than the number available, then the resulting error is called a *round-off error* – the exact value must be *rounded off* to a certain number of significant figures. Since we happen to know the result of dividing 1 by 7, we know that 0.143 contains a round-off error of about 0.00014. But when, as is usually the case, we do not know the exact value of an answer obtained using some calculating device, we assume the worst. For example, suppose that the result of some calculation worked to three significant figures is 0.527. In fact, the answer may be anything between 0.52650... and 0.52749.... In other words, 0.527 may be in error due to round-off by ±0.0005, i.e. by five units in the position after the least significant figure which is retained.

Round-off errors are not limited only to decimal fractions, or to the division operation. Working to three significant figures, the results of rounding-off several calculations are shown in Table 1.1. It should be noted that the values of percentage error given in the table are themselves in error: for example, the result 4.69 for the calculation 150/32 is actually in error by

Table 1.1 Round-off errors.

Calculation	Exact result	Rounded result	Error (%)
150/32	4.6875	4.69	0.05
1000/999	1.001	1.00	−0.1
1000/6	166.6	167	0.2
234 × 789	184 626	185 000	0.2
1000 + 1	1001	1000	−0.1

0.053%. However, for the present purpose it is adequate to round this result to 0.05%.

Modern calculators work with at least eight significant figures. Round-off error is therefore not normally a serious problem. However, there are some situations in which it can be significant. One such situation arises when two nearly equal numbers are subtracted. The round-off error can then be relatively large compared with the result. For example, working to three significant figures, the result of the calculation

$$(0.527 \pm 0.0005) - (0.517 \pm 0.0005)$$

is 0.010 ± 0.001, since in the worst possible case the round-off errors can be cumulative. Thus the *relative* round-off error has grown from 0.1% to 10%. There is nothing that can be done to avoid this problem, except to use more significant figures. If that is not possible, then loss of precision is inevitable.

If a calculation involves many arithmetic operations, then the cumulative effect of round-off errors can become serious even if the foregoing situation does not arise. The *stability* of some numerical procedures will be discussed in later sections, but it should be mentioned here that certain procedures are unstable: they amplify errors (from whatever source, including round-off errors) without bound, and therefore cannot be used. It should not be assumed that the use of, say, 12 significant figures in a calculation will result in an answer with the same precision.

Another type of error is illustrated by the evaluation of the sum of an infinite series. Consider

$$S = \frac{1}{2} + \frac{1}{4} + \frac{1}{8} + \frac{1}{16} + \cdots + \frac{1}{2^N} + \cdots \qquad (1.8)$$

Since we cannot include all of the infinite number of terms in the series, and since we see that successive terms are contributing progressively smaller and smaller amounts to S, we conclude that we can *truncate* the series after some finite number of terms (N) and so obtain an approximation to S, which we might denote by S_N. The error $S - S_N$ is called the *truncation error* in S_N.

It is important – and this will be discussed in more detail in later sections – that it should be possible to reduce the truncation error by taking more terms

Table 1.2 Reduction in truncation error in the evaluation of the sum of the series (1.8)

N	S_N	Error in S_N
3	0.875	0.125
5	0.968 75	0.031 25
10	0.999 02	0.000 98
15	0.999 97	0.000 03

in the series, i.e. by increasing N. In the case of (1.7) it is easy to believe by inspection of the general term in the series (and it can be proven) that successive values of S_N converge to a limit which is S, the sum of the infinite series. Table 1.2 shows the values of S_N and the corresponding error (which we can calculate in this case because we know from other theory that the true value of S is 1) for several values of N. We can see that the error diminishes as N increases, and that S_N is converging to a limit. (Note, however, that we have not *proved* this.)

A further illustration of round-off and truncation errors is provided by the two methods for the solution of quadratic equations described in Section 1.1. Consider the problem of finding the *smaller* root of

$$x^2 - 25x + 1 = 0 \tag{1.9}$$

Equation (1.2) yields

$$x = \{25 - \sqrt{(625 - 4)}\}/2$$

which, if we are again limited to three significant figures, becomes

$$\begin{aligned} x &= (25 - \sqrt{621})/2 \\ &= (25 - 24.9)/2 \\ &= 0.05 \end{aligned} \tag{1.10}$$

However, if we are permitted six significant figures we obtain

$$\begin{aligned} x &= (25 - 24.9199)/2 \\ &= 0.04005 \end{aligned} \tag{1.11}$$

The value given by the use of only three significant figures is almost 25% high; the use of six significant figures yields an answer which is about 0.04% low. The problem arises when the round-off error is comparable with the quantity being computed, which can happen, as here, when the difference between two almost equal numbers must be computed.

Now consider the problem of finding the *larger* root of (1.9) using (1.3), i.e.

$$x = (25x - 1)/x \tag{1.12}$$

NON-DIMENSIONAL EQUATIONS 11

Table 1.3 The truncation error in an iterative process

x	RHS of (1.12)	Truncation error
20	24.95	0.0099
24.95	24.959 92	0.000 02
24.959 92	24.959 94	0.000 00
24.959 94	24.959 94	0.000 00

If we were to guess that the solution was 20, the repetitive use of (1.12) in the manner discussed in Section 1.1 would lead to the results shown in Table 1.3. The errors listed in this table were calculated from the true solution which, to seven significant figures, is 24.959 94. We call these errors *truncation errors*, because they are the errors present in the latest value of x, assuming that the process of substitution and evaluation is truncated at that point. We note that the truncation error diminishes after each additional iteration – clearly, a necessary feature of an iterative process.

1.7 Non-dimensional equations

As part of the development of a mathematical model it is desirable, wherever appropriate, to make use of non-dimensional, rather than dimensional, equations. There are two reasons for this.

In the first place, the careful use of scale factors will result in many, if not all, of the problem variables being reduced to unit order of magnitude. This has the practical consequence of making overflow and underflow* less likely when a computer is used to implement the algorithm. It also often aids in an assessment of the relative significance of the various terms in the equations, and in an interpretation of the results of the calculations.

In the second place, and from a practical point of view perhaps more importantly, the number of independent parameters is generally reduced. As a consequence, if a parametric study has to be made over a range of conditions (i.e. over a range of values of the independent parameters), then the number of solutions which must be computed is correspondingly reduced.

Consider, for example, the problem of Section 1.5. If we limit our attention to a sphere, for which the cross-sectional area $A = \pi L^2/4$, where L is the diameter of the sphere, and for which the mass m can be written $\rho \pi L^3 / 6$, (1.6) becomes

$$\frac{dv}{dt} = \left(1 - \frac{\rho_1}{\rho}\right) g - \frac{3\rho_1}{4\rho L} f(\text{Re}) \, v^2 \qquad (1.13)$$

---

* Numbers becoming, respectively, greater or smaller in magnitude than the greatest and smallest permitted by the computer being used.

Equation (1.13) shows that the velocity of the falling sphere, as a function of time, depends on four quantities: the density ratio (ρ_1/ρ), the diameter (L) of the sphere, the gravitational acceleration (g) and the Reynolds Number (Re).

If we introduce the quantities $V = v/\sqrt{(Lg)}$ and $\tau = t/\sqrt{(L/g)}$, which are readily shown to be non-dimensional, and the quantity $\alpha = \rho_1/\rho$, which is already non-dimensional, (1.13) becomes

$$\frac{dV}{d\tau} = (1 - \alpha) - \frac{3}{4}\alpha f(\text{Re})\, v^2 \qquad (1.14)$$

Finally, noting that

$$\text{Re} = \frac{\rho_1 v L}{\mu} = \frac{\rho_1 \sqrt{(gL^3)}}{\mu} V = \beta V \quad \text{(say)}$$

where β is a new non-dimensional number, we may write (1.14) as

$$\frac{dV}{d\tau} = (1 - \alpha) - \frac{3}{4}\alpha f(\beta V)\, V^2 \qquad (1.15)$$

showing that the non-dimensional velocity V, as a function of non-dimensional time τ, now depends on only the two quantities α and β. To cover all possible problems, solutions need to be computed for ranges of values of just two parameters, and not four as in the original problem (1.13). The initial condition (1.7) becomes

$$V = 0 \quad \text{at } \tau = 0 \qquad (1.16)$$

This technique can be applied to many problems, with similar labour-saving results, and should be employed whenever possible.

1.8 The use of computers

Few, if any, of the methods described in this book can be implemented realistically using only pencil and paper. Some sort of aid to calculation will be required. Hand-held electronic calculators, the simplest of which are available at very low cost, are essential to a science or engineering student (as the slide rule was a few years ago). They enable the student to test and use many numerical methods without demanding an excessive amount of labour.

However, some of the methods are only suitable for use on a computer or programmable calculator. Calculators which can store a program, and even read a program which was previously written onto a magnetic card, can now be obtained. Home computers, with a high-level programming language, can be purchased at a cost comparable with that of a television set. These devices are equivalent, in all essential respects, to a multi-million dollar high-speed digital computer. The differences are of degree, rather than of kind: in the number of operations they can perform per second, in the number of program steps and memory locations they can accommodate, and in the level of the programming language they accept.

It is therefore assumed that all students using this book will have access to some sort of programmable calculator or computer in their homes, as well as at their university or college. This is really necessary because it is essential for students to write, and test, computer programs for the methods described. It is only by actually applying the various methods that they will understand and become fully familiar with them. This will, in turn, encourage them to use these methods in the other subjects that they are studying. It should be remembered that these methods are not an end in themselves – they are an aid to finding the solution of problems in engineering or science.

On the other hand, it should also be remembered that computer time costs money, and it is important to program economically. However, efficient programming is an art. Students should make opportunities to develop this art during their training. If they can have access to a programmable calculator, so much the better – the programming techniques used for them are similar to those employed on 'real' computers.

Numerical methods have been employed for many years, but it has only been since World War II, and since the advent of the digital computer, that they have become the powerful tool of science that they are today. Considering the number of computers now in use, it is perhaps hard to realize that they have been in existence for barely one human generation, and that in that time there have been four generations of computers: from the first electromechanical machines through electronic computers using vacuum tubes and more modern solid state devices, to the 'super-computers' of today (which are really many computers operating in parallel). The speed of super-computers is measured in 'megaflops', or millions of floating point operations per second. A speed of the order of 100 megaflops is now (late-1980s) available, compared with a predicted speed of the order of 1 flop in 1937! Calculations can therefore now be performed at more than 10^8 times the speed of the earliest computers, and perhaps 10^9 times the speed of hand-calculations. Put another way, it would take a human with a hand-held calculator, performing one operation per second without error, about 15 years to perform the same number of calculations that a modern computer can do in one second. The computer is, indeed, a powerful tool.

2
The solution of equations

2.1 Introduction

There are only a few types of equations which can be solved by simple, direct methods in a finite (and predictable) number of operations. These include the linear equation

$$ax + b = 0 \tag{2.1}$$

and the quadratic equation

$$ax^2 + bx + c = 0 \tag{2.2}$$

for which the solutions are, of course, well known.

The majority of equations encountered in engineering and scientific applications are of types which do not possess a direct solution procedure. There are two general classes of equations. The first, *algebraic* equations, contains only those functions which can be constructed using the mathematical operations of addition, multiplication and involution (the raising of a number to a power) together with the respective inverse operations. Equations (2.1) and (2.2) are simple algebraic equations. If a small change is made in (2.2), viz.

$$ax^{2.1} + bx + c = 0 \tag{2.3}$$

we obtain an equation which is still algebraic, but which is not so readily solved.

The second class of equations is called *transcendental*, and can be defined, somewhat negatively, as consisting of equations containing functions which are not all algebraic. The commonest transcendental functions are the trigonometric, exponential and logarithmic functions. Thus

$$ax^2 + b \sin x + c = 0 \tag{2.4}$$

is a transcendental equation.

Equations may also be *differential* or *integral*, but we are not concerned in this chapter with these types.

When *direct* methods of solution are not available, *iterative* techniques must be used. In an iterative technique, an initial estimate of the solution is obtained somehow, and successive improvements to this estimate are then constructed. Iterative techniques are inherently approximate. We can (in principle, at least) make the error as small as we wish, but we can never (except in freak circumstances) find the exact solution. Direct methods, on the other hand, do lead to the exact solution (apart from the possible effects of round-off errors).

2.2 Location of initial estimates

In an iterative solution procedure, it is necessary to have an initial estimate of the solution (or solutions). This may be accomplished in several ways.

Sometimes an equation can be modified to a form which can be solved directly. For example, we might expect the solutions of (2.2) to be approximations to the solutions of (2.3). As a particular case, the solutions of

$$x^2 - 4x + 2 = 0 \tag{2.5}$$

are $2 \pm \sqrt{2}$ or, approximately, 3.414 and 0.586; and it can be easily shown (after studying this chapter!) that the solutions of

$$x^{2.1} - 4x + 2 = 0 \tag{2.6}$$

are 2.985 and 0.579. The solutions of (2.5) are therefore approximately the same as the solutions of (2.6), and could be expected to make satisfactory estimates for those solutions.

Another example is provided by the polynomial equation

$$x^5 - x - 500 = 0 \tag{2.7}$$

It is apparent that the largest solution (in absolute value) of this equation can be estimated by neglecting the middle term, because if there is a solution S somewhat greater in magnitude than unity (and it certainly looks as though there should be), then S will be small compared with S^5. Thus we are led to the estimate $x \approx 500^{1/5} = 3.466$. One solution of (2.7) turns out to be 3.471. Neglecting the middle term on the right-hand side of the equation – having convinced ourselves that it is small compared with the first term – has allowed us to simplify the equation to a point where it can be solved readily, and the solution of the simplified equation can then be used as a first estimate of the solution of the original equation. (We shall return to a consideration of how to solve polynomial equations in Section 2.11.)

When the solutions cannot be estimated in this or some similar manner, we can locate them approximately by evaluating the function for a range of values of x or, equivalently, by graphical means. Suppose we wish to solve

$$f(x) = x^2 - \sin x - 5 = 0 \tag{2.8}$$

It is reasonable to suppose that since $-1 \leq \sin x \leq 1$, there will be solutions of (2.8) at approximately $\sqrt{5}$, i.e. at about ± 2.2. We verify this by evaluating

Table 2.1 Values of $f(x) = x^2 - \sin x - 5$.

x	$f(x)$
-4	10.2
-3	4.1
-2	-0.1
$\vdots$	$\vdots$
2	-1.9
3	3.9
4	11.8

$f(x)$ for several values of x, as shown in Table 2.1. Since $f(x)$ changes sign between -3 and -2, and again between 2 and 3, it must pass through zero in these ranges, i.e. there is some value of x within each of these ranges that makes $f(x)$ zero.

The solutions can be estimated more accurately graphically. Equation (2.8) can be written

$$x^2 - 5 = \sin x \qquad (2.9)$$

We can plot each side of (2.9) against x and locate the solutions at the intersections of the graphs. This is shown in Figure 2.1, from which it can be

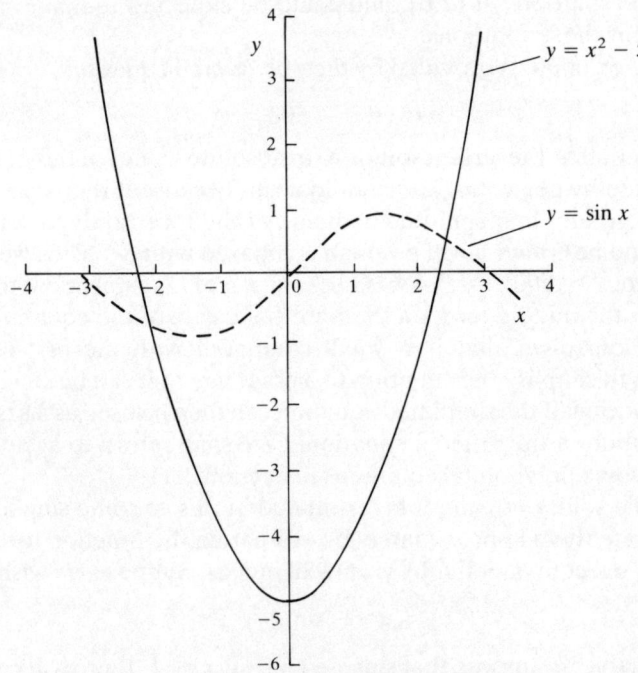

Figure 2.1 A graphical estimation of the solutions of (2.9).

LOCATION OF INITIAL ESTIMATES

```
c
        subroutine search(x,xhi,delx,found)
c
c       to locate solutions of  f(x) = 0  within a range
c       from  x  to  xhi  with an accuracy of  delx.
c
        logical found
        found = .false.
        xlo = x
        p = f(x)
    5   q = f(x+delx)
        if ( p*q .lt. 0.) then
            found = .true.
            y = x + delx
            write (*,15) x,y
   15       format(' a solution lies between ',f5.1,' and ',f5.1)
            return
        else
            x = x + delx
        endif
        if ( x .ge. xhi ) then
            write (*,25) xlo,xhi
   25       format(' no solution found between ',f5.1,' and ',f5.1)
            return
        else
            go to 5
        endif
        end
```

Figure 2.2   The Fortran listing of subroutine SEARCH.

seen that the solutions are approximately −2.1 and 2.4. These values can now be used as initial estimates for the iterative procedures discussed in later sections.

The technique of locating a solution of the continuous function $f(x) = 0$ by finding an interval within which $f(x)$ changes sign is easily implemented on a digital computer. We simply choose a starting value for x – call it x_1 – and evaluate $f(x_1)$. We then change x to x_2, and find $f(x_2)$. If $f(x)$ has changed sign, then there is a solution between x_1 and x_2. If not, there is no solution between x_1 and x_2, so we abandon this interval, choose a value x_3 greater than x_2, and compare the signs of $f(x_2)$ and $f(x_3)$. We continue in this manner until $f(x)$ does change sign. The method of choosing x_2, x_3, etc., must be systematic if it is to be accomplished on a machine.

Figure 2.2 shows the listing of a Fortran Subroutine 'search', which carries out this procedure. The successive trial values of x are, for simplicity, chosen at equal intervals of Δx (or, in the program notation, 'delx'). We need not retain the trial values of x after they have been discarded; that is, after an unsuccessful test the 'old' value of x is overwritten by $x + \Delta x$. The test for a change of sign of $f(x)$ is very simple: we ask whether $f(x) \cdot f(x + \Delta x) < 0$. If there has been a change of sign, either from negative to positive or vice versa, then the answer is 'yes'. If the answer is 'no', then the quantity $f(x) \cdot f(x + \Delta x)$ must be positive or zero. If it is positive, then there is no zero of $f(x)$ in this interval (or possibly there are two zeros here – see the discussion of Figure 2.4, below) and we continue the search to the next higher interval. If it is zero, then either x or $x + \Delta x$ happens to be a solution of the equation $f(x) = 0$, and it is easy to check which one is.

It has been assumed here that the function $f(x)$ has been defined in a separate function subprogram. This keeps the subroutine 'search' quite general. Alternatively, $f(x)$ could be defined in a function statement within the subroutine.

There are four parameters in the subroutine: the lower and upper limits of the range over which the search is to be conducted, the increment Δx and a logical variable 'found', which has the value '.true.' if a change of sign has been encountered and is '.false.' otherwise. The last parameter is necessary to ensure that the solution procedure we ultimately adopt does not select the upper limit of the range of search as a first estimate if a return to the main program is effected *without* a sign change being encountered. Upon exit from the subroutine after a successful 'search', the first parameter contains the value of x at the beginning of the interval within which the change of sign occurs.

Figure 2.3 shows the listing of a simple main program which calls 'search'. In this example, estimates of the solutions of the equation

$$\sin x + x/5 - 1 = 0 \qquad (2.10)$$

which lie between 0 and 20 have been found, using a search increment Δx of 0.1.

```
      c
              logical found
      c
              x = 0.
              xhi = 20.
        5     call search (x, xhi, 0.1, found)
              if( .not. found) then
                  write (*,15)
       15         format(' upper limit of search range reached')
                  stop
              else
                  x = x + 0.1
                  go to 5
              endif
              end

              function f(x)
              f = sin(x) + x/5. - 1.
              return
              end
```

Figure 2.3 Use of the subroutine SEARCH.

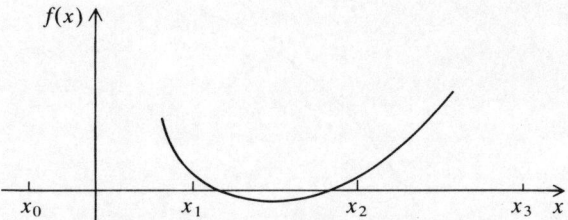

Figure 2.4 Failure of the search procedure.

The output from the program of Figures 2.2 and 2.3 appears as follows:

```
a solution lies between   .9 and   1.0
a solution lies between   2.6 and   2.7
a solution lies between   6.0 and   6.1
no solution found between   6.1 and   20.0
upper limit of search range encountered
```

How large or small should the search increment Δx be? It is not possible to give a general rule for this. If Δx is very small, then the 'approximate' estimate of the solution will be quite accurate, and the solution procedure itself will have less work to do. However, if the search range is large, the increment small and the function $f(x)$ very complicated, then the search procedure can become time-consuming. On the other hand, if the increment Δx is too large, the situation shown in Figure 2.4 becomes more likely to happen, in which the search procedure has jumped over the two solutions lying between x_1 and x_2.

At the price of increasing the complexity of subroutine 'search', it can be refined to guard against such possibilities. For example, the size of the increment Δx could be decreased somewhat – say by a factor of 2 – if $|f(x)|$ approaches zero, and allowed to increase as $|f(x)|$ grows. Another modification which could be included is a section to detect whether either p or q [i.e. either $f(x)$ or $f(x + \Delta x)$] happens to be zero or very close to it, in which case x or $x + \Delta x$, respectively, would be the actual solution. Such refinements are left for the student to explore; the problems at the end of the chapter include some examples chosen to lead to such pathological situations.

2.3 Interval halving

One of the simplest, yet most effective, methods for the solution of an equation $f(x) = 0$ is based on finding, using a procedure such as 'search', an interval over which $f(x)$ changes sign. Figure 2.5 shows the method, which is known as the method of *interval halving*.

THE SOLUTION OF EQUATIONS

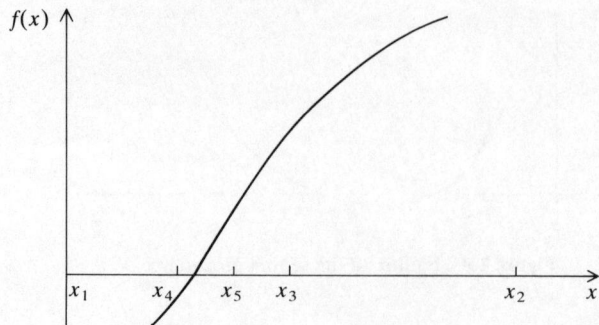

Figure 2.5 The method of interval halving.

Suppose 'search' has located the values x_1 and x_2 which define an interval within which $f(x)$ changes sign. The value of x at the mid-point of this interval is computed: $x_3 = (x_1 + x_2)/2$. The sign of $f(x_3)$ is now compared with that of the function at one end or the other of the interval – with that at x_1, say. If the two signs are different, as in Figure 2.5, then the solution lies between x_1 and x_3. The value x_2 is discarded, and the process is repeated in the new interval (x_1, x_3). With $x_4 = (x_1 + x_3)/2$, the solution is found to lie between x_4 and x_3. In this manner, the size of the interval within which the solution lies is repeatedly halved. At each step we can assume that the latest and best estimate of the solution is the mid-point of the current interval. The maximum possible error in this estimate of the solution is half the current interval size, and is itself halved with each iteration. If the interval in 'search' [i.e. $(x_2 - x_1)$] is denoted by Δx, and if $x_1 + \Delta x/2$ is adopted as the first estimate of the solution, then the maximum possible error is $\Delta x/2$. Hence, after N interval-halving operations the maximum possible error is $(\Delta x/2) \cdot 2^{-N} = \Delta x/2^{N+1}$. For example, with $\Delta x = 0.1$ the error after 10 iterations is $0.1/2^{11} = 5 \times 10^{-5}$. An extra significant figure is gained, on the average, after every three or four iterations.

The outstanding advantage of the method of interval halving is that it is *guaranteed to work for all continuous functions*, provided that it is supplied with the starting values needed on either side of the solution. This can be said about few, if any, of the other methods in common use. A second advantage is that we can calculate *in advance* how many iterations will be needed to reduce the error to a specified fraction of the initial error. If the initial and required errors are E_i and E_r, respectively, then the smallest integer larger than

$$N = \log_2 (E_i/E_r) = 3.32 \log_{10} (E_i/E_r) \qquad (2.11)$$

is the number of iterations necessary.

As an example of interval halving, consider the determination of the first positive root of

$$f(x) = \cos x \cosh x + 1 = 0$$

– an equation which is relevant to the theory of a vibrating cantilever.

The first stage is to locate an interval within which the solution is located. Starting at $x = 0$, and using $\Delta x = 0.4$, a search for a change in the sign of $f(x)$ yields

x	$f(x)$
0	2.0000
0.4	1.9957
0.8	1.9318
1.2	1.6561
1.6	0.9247
2.0	−0.5656

showing that a solution lies between $x = 1.6$ and $x = 2.0$.

Denoting by x_l, x_r and x_m the values of x at the left end, right end and mid-point of the interval within which we now know the solution to lie, we can compute the following values:

x_l	x_m	x_r	$f(x_l)$	$f(x_m)$	$f(x_r)$
1.6	1.8	2.0	0.9247	0.2940	−0.5656
1.8	1.9	2.0	0.2940	−0.1049	−0.5656
1.8	1.85	1.9	0.2940	0.1020	−0.1049
1.85	1.875	1.9	0.1020	0.0004	−0.1049
1.875	1.8875	1.9	0.0004	−0.0518	−0.1049
1.875	1.8813	1.8875	0.0004	−0.0255	−0.0518
1.875	1.8781	1.8813	0.0004	−0.0125	−0.0255
1.875	1.8766	1.8781	0.0004	−0.0060	−0.0125
1.875	1.8758	1.8766	0.0004	−0.0028	−0.0060
1.875	1.8754	1.8758	0.0004	−0.0012	−0.0028
1.875	1.8752	1.8754	0.0004	−0.0004	−0.0012

It would not normally be necessary to record all this information; it has only been given here to show the progress of the calculations. In particular, the *values* of $f(x)$ are not needed; it is sufficient to record the *sign* of the values.

After the interval containing the solution has been located, successive iterations reduce the size of the interval by a factor of 2. After ten iterations, therefore, the interval has been reduced to $0.4/2^{10} = 0.00039\ldots$. The solution lies between 1.875 and 1.87539.... If we were to stop the iterations at this stage, we would accept the mid-point of *this* interval, 1.8752, as the best estimate of the solution. A further three iterations would yield 1.8751, which is correct to five significant figures.

A flowchart for a computer program using interval halving is shown in

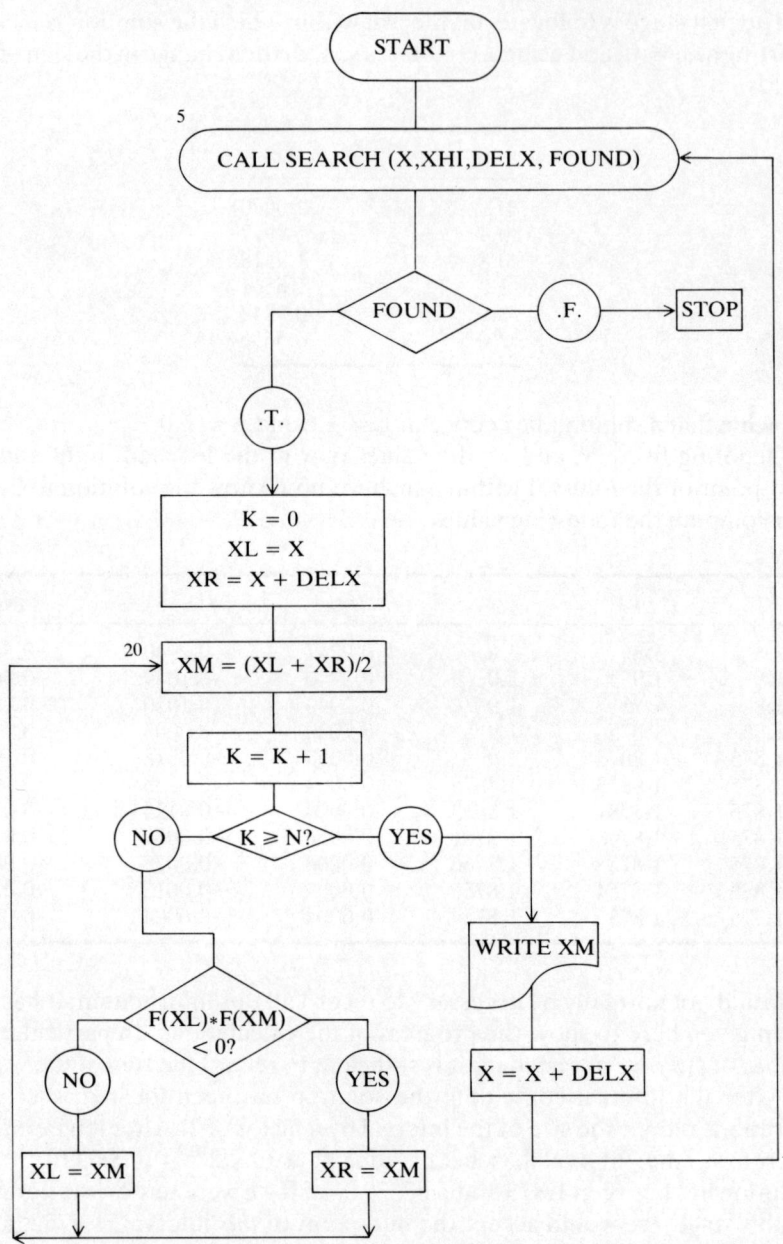

Figure 2.6 A flowchart for the solution of $f(x) = 0$ using interval halving.

```
c
        logical found
c
        n = 10
        x = 0.0
        delx = 0.1
        xhi = 20.
c
    5   call search (x, xhi, delx, found)
c
        if ( .not. found) then
            write (*,10)
   10       format (' upper limit of search range reached')
            stop
        endif
c
   15   k = 0
        xl = x
        xr = x + delx
   20   xm = (xl + xr) / 2.
        k = k + 1
c
        if ( k .ge. n ) then
            write (*,25) xm
   25       format (' the solution is ',f8.4/)
            x = x + delx
            go to 5
        endif
c
        if ( abs( f(xm) ) .lt. 1.0e-05 ) then
            write (*,25) xm
            x = x + delx
            go to 5
        endif
c
        if ( f(xl)*f(xm) .gt .0.0) then
            xl = xm
            go to 20
        endif
c
   35   xr = xm
        go to 20
        end

        function f(x)
        f = sin(x) + x/5. - 1.
        return
        end
```

Figure 2.7 The Fortran listing of a program for the method of interval halving.

Figure 2.6. As in the case of 'search', only the values of X at the left- and right-hand ends of the current interval are required. These are called XL and XR, and one or the other of them is modified at each iteration. The mid-point of the interval is denoted by XM, and the half-interval in which the solution lies is identified by examining the sign of F(XL) * F(XM). The process is terminated after N iterations, where N has to be calculated using (2.11) in accordance with the desired accuracy and specified at the start of the program (or read as a variable to allow the accuracy to be changed without having to recompile the program). K is the iteration counter: a variable with a value equal to the number of completed iterations.

A program based on the flowchart is given in Figure 2.7. The subroutine 'search' is used to find the starting values. The program will find all of the solutions of $f(x) = 0$ (which is defined in a separate function subprogram) which lie between the limits 'xa' and 'xb'. In Figure 2.7 the results of using the program to solve (2.10), with 'xa' = 0 and 'xb' = 20, are also reproduced.

A section has been included in the program to detect and respond if XM happens to be the solution of the equation. This is the section commencing

```
if ( abs( f(xm) ) .lt. 1.0e-5 ) then
```

Notice that it would not be satisfactory to ask whether $f(x_m)$ was exactly zero – the chances of a real variable being exactly zero are very small. It is necessary to ask whether the absolute value of $f(x_m)$ is 'small' – here defined as being less than 10^{-5}.

A section has not been included to detect whether the lower or upper ends of the search region happen to be solutions. It is assumed that this check is now contained in the subroutine 'search'.

The output from the program in Figure 2.7 (with which the 'search' subroutine must be included) appears as follows:

```
a solution lies between    .9 and    1.0
the solution is      .9456

a solution lies between   2.6 and    2.7
the solution is     2.6530

a solution lies between   6.0 and    6.1
the solution is     6.0679

no solution found between  6.1 and   20.0
upper limit of search range reached
```

2.4 Simple iteration

A large class of iterative methods for the solution of

$$f(x) = 0 \quad (2.12)$$

is based upon rewriting (2.12) in an equivalent form:

$$x = F(x) \quad (2.13)$$

For example,

$$x^3 - 3x^2 - 3.88x + 3.192 = 0 \quad (2.14)$$

may be written

$$x = (x^3 - 3x^2 + 3.192)/3.88 \quad (2.15)$$

If an initial estimate x_0 of the solution has been found, then a sequence of values (x_n) can be computed from the relationship

$$x_{n+1} = (x_n^3 - 3x_n^2 + 3.192)/3.88 \quad (2.16)$$

SIMPLE ITERATION

Starting with $x_0 = 0.5$, for example, we find that

$$x_1 = 0.662$$
$$x_2 = 0.559$$
$$x_3 = 0.626, \text{ etc.}$$

If the iterations are continued, it will be found that successive values of x_n become increasingly close to 0.6. If we were able to retain a sufficient number of significant figures in our computations, we would find that x_n always remains slightly different from 0.6. However, we could make this difference as small as we wish (within the limits of accuracy of the computing device being used) by calculating further iterations. We say that the sequence x_n *converges* to the limit 0.6, and the process itself is said to be *convergent*.

On the other hand, if we start with $x_0 = 4$, we find that

$$x_1 = 4.9$$
$$x_2 = 13.1$$
$$x_3 = 447.1, \text{ etc.}$$

Successive terms in the sequence continue to grow rapidly in value. Now the process is said to be *divergent*.

Nevertheless, there *is* a solution of (2.14) near $x = 4$, for if we apply iteration to

$$x = 3 + 3.88/x - 3.192/x^2 \qquad (2.17)$$

which is another rearrangement of (2.14), we find that

$$x_1 = 3.770$$
$$x_2 = 3.805$$
$$x_3 = 3.799, \text{ etc.}$$

and the sequence continues to converge to the solution $x = 3.8$.

It therefore appears that a given iterative formula can be convergent to one solution, but not to another, and that an alternative iteration formula can, at least in this example, be found for the troublesome solution*.

It is also possible, in freak circumstances, for an iterative process to be neither convergent nor divergent, but to get into an endless loop. Two such possibilities are mentioned later (in Sections 2.5 and 2.6). Special precautions should be taken in computer programs to guard against such possibilities by limiting the maximum number of iterations the program can execute.

---

* In the present context, the concept of convergence would be better described by the term 'iterative convergence', to distinguish it from the concept of 'mesh size convergence' which will be discussed in the chapters on differential equations. Iterative convergence is linked with 'stability', which is also of relevance to the solution of differential equations: a solution procedure is said to be stable if errors do not grow without bound as the solution progresses. The double meaning of the word 'convergence' is unfortunate, but well-entrenched.

2.5 Convergence

To consider these results a little more formally, we return to the general form (2.13), for which the iteration formula is

$$x_{n+1} = F(x_n) \tag{2.18}$$

Let S denote the solution [i.e. that value of x which satisfies (2.13)], and let the error in x_n be e_n, i.e.

$$e_n = x_n - S \tag{2.19}$$

Substitution of (2.19) into (2.18) yields

$$S + e_{n+1} = F(S + e_n)$$

The right-hand side can be expanded in a Taylor series about S:

$$S + e_{n+1} = F(S) + e_n F'(S) + (e_n^2/2!)F''(S) + \cdots$$

Since S is the solution of (2.13), $S = F(S)$ and hence

$$e_{n+1} = e_n F'(S) + (e_n^2/2!)F''(S) + \cdots \tag{2.20}$$

If e_n is sufficiently small, and provided that $F'(S) \neq 0$, then

$$e_{n+1} \approx e_n F'(S) \tag{2.21}$$

For (2.18) to be convergent, it is necessary that

$$|e_{n+1}| < |e_n|$$

Equation (2.21) shows that this will be achieved if

$$|F'(S)| < 1 \tag{2.22}$$

We have thus found a *convergence criterion* for the iteration process (2.18); a condition which must be satisfied if the process is to converge.

This analysis also tells us something about the rate of convergence – about how quickly x_n approaches S and e_n approaches zero. We say that this process is of the *first order* because, by (2.21), the error in any estimate of the solution is (approximately) proportional to the *first* power of the error in the previous estimate. It is reduced, at each step, by a factor approximately equal to $F'(S)$.

In order to derive (2.21) from (2.20) we assumed that e_n was 'sufficiently small' to enable the quadratic and higher-order terms in e_n to be neglected. This means that the convergence criterion has only been proven for initial estimates which are 'sufficiently' close to S. It does not mean that the process will necessarily diverge with a bad first guess – it may do so, or it may not. However, we can guarantee that it will converge, if (2.22) is satisfied, *provided the initial estimate x_0 is good enough*.

Of course, there is one problem with (2.22): to apply the test it seems that we need to know the solution S before we start! However, if we have located an interval within which the solution lies, then it will be sufficient to consider the *maximum* magnitude of $F'(x)$ within that interval. If this is less than 1, then (2.22) will be satisfied for *all* x in the interval, and therefore, in particular, for $x = S$.

Consider (2.16) again. Suppose we have discovered somehow that solutions lie in the intervals (0, 1) and (3, 4). In this case

$$F'(x) = (3x^2 - 6x)/3.88 \qquad (2.23)$$

The maximum magnitude of $F'(x)$ within the first interval is 0.773, occurring when $x = 1$. Thus (2.18) will be convergent for *any* x in the interval. However, for $3 \leq x \leq 4$, the *minimum* value of $F'(x)$ is 2.320 (when $x = 3$) and hence (2.18) will not converge *to the solution within that interval* for any x in the interval. This emphasis is necessary because it often happens that a divergent iterative process will lead to a second or later estimate which lies within the range of convergence of a solution other than the one currently being sought.

What will happen if $|F'(S)|$ is exactly equal to 1? Equation (2.21) suggests that the error will remain constant. But (2.21) is only an approximation. So consider instead (2.20). It can be seen that, even if the iterations start to converge, a stage will eventually be reached when the second- and higher-order terms are smaller than the minimum precision of the computer or calculator being used. When that occurs, (2.21) becomes an exact equation, and e_{n+1} is then equal to e_n. Because of what might be called hardware limitations – the inability to carry an infinite number of significant figures – the iteration process will continue indefinitely, neither converging (past a certain stage) nor diverging.

As an example, consider

$$x_{n+1} = F(x_n) = (3 - x_n^2)/2$$

the solution of which is $S = 1$, and for which $|F'(S)| = 1$. Starting with, say, $x_0 = 0.5$, the process appears to be converging:

n	0	1	2	3	4	5
x_n	0.5	1.375	0.555	1.346	0.594	1.324

But the convergence is very slow. After 100 iterations (on a 16-bit microcomputer), the solution looks like this:

n	101	102	103	104	105	106
x_n	1.128	0.864	1.127	0.865	1.126	0.866

And after 30 000 more iterations, the solution becomes:

n	30 001	30 002	30 003	30 004
x_n	1.008 165	0.991 802 0	1.008 164	0.991 802 3

The iterations are still converging, but progress is painfully slow.

Eventually, however, a stage is reached when no further progress can be made:

n	? + 1	? + 2	? + 3	? + 4
x_n	1.005 085	0.994 902 3	1.005 085	0.994 902 3

The second and higher order terms in (2.20) are smaller than the computer can accommodate, and successive estimates of S are no longer improving. The process would continue indefinitely without further progress. In the computer program it is necessary to count the number of iterations which have been completed, and to terminate the procedure when some upper limit has been reached, in order to guard against such an eventuality.

The convergence criterion (2.22) possesses a simple geometric interpretation. Figures 2.8a and b show functions for which $|F'(S)| < 1$ for a range of values of x. The solution of $x = F(x)$ lies at the point of intersection of the curve of $F(x)$ versus x and the straight line x versus x. The sequence of estimates $x_1, x_2, \ldots$ is clearly approaching this point. In Figures 2.8c and d, on the other hand, for which $|F'(S)| > 1$, the iteration process is diverging: successive estimates lie increasingly far from the solution.

Equation (2.21) shows that the error is reduced, with each iteration, by a factor of about $F'(S)$. The rate of convergence therefore depends on the particular problem, and it will not be possible – as was the case with interval halving – to determine in advance the number of iterations necessary to achieve a desired accuracy.

The approach to convergence must therefore be monitored, and the iterations be terminated when the latest estimate of the solution is sufficiently close to the true solution. This can be assessed by comparing two successive estimates. Since [provided (2.22) is satisfied] the process is convergent, the limit which successive estimates approach must be the solution, and the difference between successive estimates should therefore give an indication of the difference between the latest estimate and the solution.

From (2.21) it follows that

$$x_{n+1} - S \approx (x_n - S)F'(S)$$
$$= x_n + x_n\{F'(S) - 1\} - S F'(S)$$

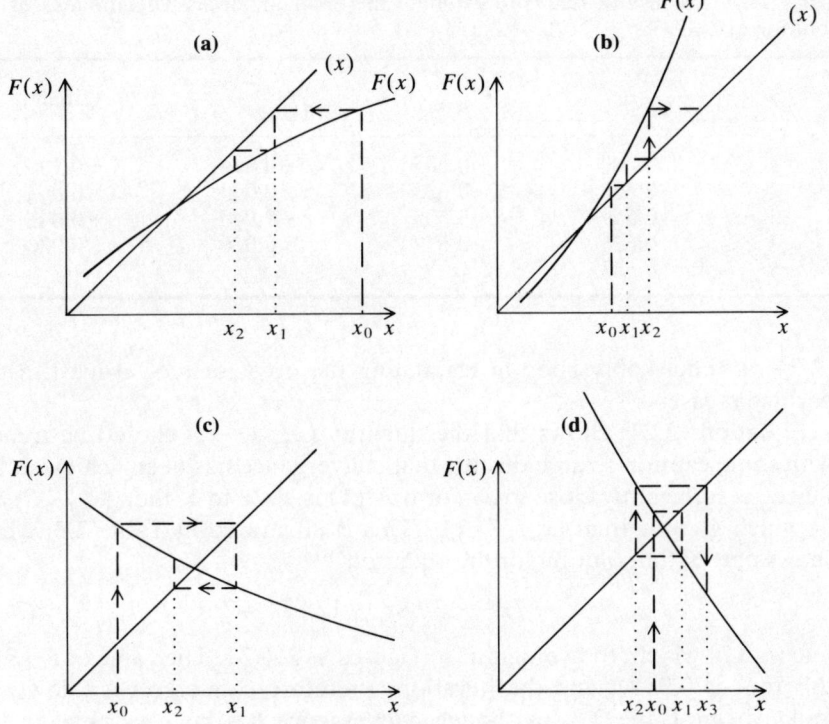

Figure 2.8 (a & b) convergent processes; (c & d) divergent processes.

Therefore
$$x_{n+1} \approx (x_n - S)\{F'(S) - 1\}$$
Finally,
$$x_n - S = e_n \approx \frac{x_{n+1} - x_n}{F'(S) - 1} \tag{2.24}$$

This equation allows us to calculate – or at least to estimate – the accuracy of x_n.

Unfortunately, $F'(S)$ cannot be evaluated until S is known. It can, however, be approximated by $F'(x_{n+1})$; (2.24) can then be replaced by

$$x_n - S = e_n \approx \frac{x_{n+1} - x_n}{F'(x_{n+1}) - 1}$$

the right-hand side of which can be used to assess the accuracy of x_n. An example of this is shown in Table 2.2, which gives several steps in the solution of (2.14) using the iteration formula (2.16). The results show that

Table 2.2 The use of (2.24) to estimate the error in successive estimates of the solution of $x^3 - 3x^2 - 3.88x + 3.192 = 0$.

n	x_n	$F'(x_{n+1})$	$\dfrac{x_{n+1} - x_n}{F'(x_{n+1}) - 1}$	Actual e_n
1	0.500	−0.685	−0.096	−0.100
2	0.662	−0.623	0.063	0.062
3	0.559	−0.665	−0.040	−0.041
4	0.626	−0.639	0.026	0.026
5	0.583			

(2.24) is remarkably good at estimating the error in x_n – at least in this particular case.

Equation (2.24) shows that the quantity $(x_{n+1} - x_n)$ should be treated with some caution as an indicator that convergence has been achieved, i.e. that x_n is sufficiently close to S. For if $F'(S)$ is close to 1, then $(x_n - S)$ can be much greater than $(x_{n+1} - x_n)$. This is illustrated in Table 2.3, which shows part of the solution for the equation

$$x^3 - 3x^2 - 3.88x + 12.824 = 0 \qquad (2.25)$$

one root of which, to five significant figures, is 2.4937. The value of $F'(S)$ at this root is 0.9519, and the iterations therefore converge very slowly. It looks, from Table 2.3, as though convergence has been reached to five significant figures after iteration 32, because two successive estimates are the same to this accuracy. In fact, the estimate x_{32} is not even correct in its fourth significant figure. Table 2.3 also shows that (2.24), while providing an indication of the error e_n, is not always as accurate as Table 2.2 would suggest.

It should be noted that in Table 2.3, as elsewhere in this book, the calculations were performed using a greater number of significant digits than

Table 2.3 The use of (2.24) to estimate the error in successive estimates of the solution of $x^3 - 3x^2 - 3.88x + 12.824 = 0$.

n	x_n	$F'(x_{n+1})$	$\dfrac{x_{n+1} - x_n}{F'(x_{n+1}) - 1}$	Actual e_n
1	2.5	0.9659	0.007 56	0.006 31
2	2.4997	0.9653	0.007 18	0.006 06
3	2.4995			
⋮	⋮	⋮	⋮	⋮
31	2.4953	0.9555	0.001 70	0.001 64
32	2.4952	0.9553	0.001 62	0.001 56
33	2.4952			

the number quoted in the table. For example, x_2 is not exactly 2.4997; it is 2.499 742 268..., and it was the latter value which was used to obtain $x_3 = 2.499\,493\,247\ldots$. For this reason, if the calculations were to be repeated with the retention of a different number of significant digits, then very slightly different answers might be obtained. Similarly, some of the computer programs presented here were run on a 16-bit microcomputer. If the same programs were to be run on a mainframe machine of larger word size, different results could be achieved. It would, perhaps, be desirable if students experienced this – it would emphasize the need to consider the accuracy of their results, and caution them against accepting all the digits their calculators display.

A fundamental question now arises: how should the iterations be terminated? When is x_n 'sufficiently' close to S? There are two tests which can be used.

In the first place (2.24), while not perfect, is fairly good – and gets better as x_n approaches S. After 60 iterations, for example, the estimate of the solution of (2.25) is actually in error by 0.000 386, while (2.24) predicts an error of 0.000 389. This is, of course, because $F'(x_{n+1})$ provides progressively better estimates of $F'(S)$. The iteration process can thus be terminated whenever $(x_{n+1} - x_n)/\{F'(x_{n+1}) - 1\}$ is 'sufficiently small'. This, in turn depends on the magnitude of x_{n+1} (or, more properly, of S) itself – the *relative* error should always be examined. A suitable test would therefore be

$$\left| \frac{x_{n+1} - x_n}{F'(x_{n+1}) - 1} \right| < \varepsilon |x_{n+1}|$$

where ε is a small number chosen as an acceptable compromise between accuracy and computing cost. A typical value for ε is 10^{-4}.

The second test is simply to substitute the latest x_n into the original equation, and to compare the imbalance with, say, the sum of all positive terms. Thus for (2.25), $x_{31} = 2.495\,324\,52$; the left-hand side of the equation becomes $-0.000\,272$, compared with $x_{31}^3 + 12.824 = 28.361$. It could therefore be said that the equation has been satisfied to within about $10^{-3}\%$ which, depending on the purpose for which the equation is being solved, might be considered good enough. If not, a further 30 iterations leads to $x_{61} = 2.494\,071\,33$, which satisfies (2.25) to within about $10^{-8}\%$.

It must be emphasized that these two tests of accuracy are not the same – they measure different characteristics. Thus while x_{31} causes (2.25) to be satisfied to $10^{-3}\%$ of the sum of all positive terms in the equation and x_{61} satisfies it to $10^{-8}\%$, these two estimates of S differ from the true S by much larger amounts, viz. 0.065% and 0.015% of S, respectively.

The choice of test depends on which is more important: knowing S accurately or having the equation satisfied accurately. As the example provided by (2.25) demonstrates, the latter does not necessarily imply the former.

2.6 Aitken's extrapolation

Because (2.24) provides at least an estimate of e_n (the error in x_n), it should be possible to use it to accelerate the convergence of an iterative solution. The method is known as Aitken's extrapolation, and we shall investigate it in connection with the solution of

$$x^{6.1} - \cos x - 10 = 0$$

Since $\cos x$ is small in comparison with 10, the solution of this equation will be approximately $10^{1/6} \approx 1.5$. A form of the equation suitable for iteration is

$$x_{n+1} = \left(\frac{\cos x_n + 10}{x_n^{2.1}}\right)^{1/4} = F(x_n)$$

The results of the first few iterations are

n	1	2	3	4	5	6
x	1.5	1.4399	1.4733	1.4544	1.4650	1.4591

and after 20 iterations the current estimate of the solution is $x = 1.461\,197$. The correct solution to six decimal places is $1.461\,198$.

We now suppose that (2.24) will give us a sufficiently good estimate of the error in x_n to justify its use as a correction to x_n. In order to use (2.24) we need to know $F'(x_n)$, which we will take as an approximation to $F'(S)$. Since the analytical expression for $F'(x)$ in this instance is quite complex, it is easier to estimate the derivative numerically. Methods for doing this will be studied in detail in Chapter 4; at this stage it is sufficient to use the obvious approximation

$$F'(x_n) \approx \frac{F(x_n) - F(x_{n-1})}{x_n - x_{n-1}} = \frac{x_{n+1} - x_n}{x_n - x_{n-1}} \quad (2.26)$$

in which, for good accuracy, x_{n-1} and x_n must be fairly close.

The following calculations can now be performed. Let

$$x_1 = 1.5$$

then

$$x_2 = F(x_1) = 1.439\,851$$
$$x_3 = F(x_2) = 1.473\,302$$

Therefore (using just the significant figures which are recorded here)

$$F'(x_2) \approx \frac{F(x_2) - F(x_1)}{x_2 - x_1}$$
$$= \frac{1.473\,302 - 1.439\,851}{1.439\,851 - 1.5}$$
$$= -0.566\,136$$

Therefore
$$e_2 \approx \frac{x_3 - x_2}{F'(x_2) - 1}$$
$$= \frac{1.473\,302 - 1.439\,851}{-1.566\,136}$$
$$= -0.021\,496$$

We thus obtain $x_2 - e_2 = 1.461\,347$ as a better estimate of S. Note that we have calculated the error in x_2, not that in x_3, because of the use of (2.26) to approximate $F'(S)$.

The calculations can be simplified by inserting the approximation for $F'(S)$ directly into (2.24), obtaining
$$e_n \approx \frac{(x_{n+1} - x_n)(x_n - x_{n-1})}{x_{n+1} - 2x_n + x_{n-1}}$$
and hence a better estimate of S is
$$x_n - e_n = \frac{x_{n+1}x_{n-1} - x_n^2}{x_{n+1} - 2x_n + x_{n-1}} = T \quad \text{(say)} \tag{2.27}$$

Thus, once three successive estimates of S have been obtained, T can be calculated by what is essentially an extrapolation process. This value can then be used as the first of another set of three estimates, and the process continued.

If the calculations are being performed by hand (i.e. using a calculator, not a computer), they can be further simplified by setting them out in a table. This also reduces the likelihood of arithmetic error, and is always good practice where appropriate.

n	x_n	T
1	1.5	
2	1.439 851	
3	1.473 302	1.461 347
4	1.461 347	
5	1.461 114	
6	1.461 245	1.461 197

The correct solution to an accuracy of one part in 10^6 has now been obtained after only five iterations and two calculations of T (which involve a negligible computing effort), compared with 25 iterations without extrapolation.

However, another repetition of this process seems to cause trouble. Starting with 1.461 197 (exactly) we obtain

7	1.461 197	
8	1.461 197 853	
9	1.461 197 375	1.461 307 288

and see that the latest value of T is *less* accurate than the latest value of x_n.

The reason for the trouble is round-off error, in its classic form – subtracting two numbers which differ only in their few least significant digits. The details of the calculation of the third estimate of T, using x_7, x_8 and x_9, are

$$T = \frac{x_9 x_7 - x_8^2}{x_9 - 2x_8 + x_7}$$

$$= \frac{2.135\,097\,221 - 2.135\,099\,166}{1.461\,197\,375 - 2.922\,395\,706 + 1.461\,197}$$

$$= \frac{-0.000\,001\,945}{-0.000\,001\,331}$$

$$= 1.461\,307\,288$$

The two subtractions in the second line of this calculation yield results which contain only four significant figures, with a corresponding loss of precision.

This difficulty can be avoided by subtracting out the 'constant' part of each value and working only with the remainder. For example, if 1.461 is subtracted from each of the three estimates x_7, x_8 and x_9, then the remainders multiplied by 1000 (just to keep their magnitudes more manageable) and extrapolation then applied, the result is

$$\frac{(0.197)(0.197\,375) - (0.197\,853)^2}{0.197 - 2(0.197\,853) + 0.197\,375} = 0.197\,547$$

Dividing this by 1000 and adding back 1.461 yields 1.461 197 547, which is the solution correct to nine decimal places.

Great care must be exercised in using Aitken's extrapolation. It is very helpful in accelerating convergence in the early stages of a calculation, and may be used in conjunction with any first-order process; but when the process has almost converged it becomes subject to round-off error which can cause the quality of the estimate of the solution to deteriorate if precautions are not taken.

2.7 Damped simple iteration

Equation (2.12) can be rearranged in an infinite number of ways into a form suitable for iteration. It is often possible to find arrangements which will enhance the rate of convergence; it is even possible to turn a non-convergent form into one which is convergent. Since S satisfies

$$S = F(S) \tag{2.13}$$

then it will also satisfy

$$S = \lambda F(S) + (1 - \lambda)S = G(S) \quad \text{(say)} \tag{2.28}$$

where λ is any number. The iteration form of (2.28) is

DAMPED SIMPLE ITERATION

$$x_{n+1} = \lambda F(x_n) + (1 - \lambda)x_n = G(x_n) \tag{2.29}$$

The quantity λ can be regarded as a *damping* or *interpolating* factor. Equation (2.29) expresses the fact that the value taken for x_{n+1}, which is the new estimate of S, is a combination of $F(x_n)$ (the value computed by simple iteration) and the old estimate of S, which is x_n. This is a common technique for *damping*, or stabilizing, an iterative procedure which might otherwise oscillate or even diverge.

Consider (2.21), for which $G(x)$ becomes

$$G(x) = \lambda \frac{x^3 - 3x^2 + 3.192}{3.88} + (1 - \lambda)x$$

and

$$G'(x) = \lambda \frac{3x^2 - 6x}{3.88} + 1 - \lambda \tag{2.30}$$

Suppose we know that there is a solution at $S = 0.6$. Then we can calculate

$$G'(0.6) = 1 - 1.64948\,\lambda$$

If we now choose λ to be $1.64948^{-1} = 0.60625$, we can make $G'(S) = 0$ which, by (2.21), would lead to the most rapid convergence. [Note that although this appears to lead to immediate convergence by forcing e_{n+1} to be zero, this is not actually the case because (2.21) is only an approximation to (2.20), and therefore does not calculate e_{n+1} exactly.]

Unfortunately, we do *not* yet know the value of S. We only know, from some search procedure, a first approximation to S; suppose this approximation is 0.5. Then

$$G'(0.5) = 1 - 1.58\,\lambda \tag{2.31}$$

For convergence, we must have $|G'(S)| < 1$ which means, in practice, $|G'(x)| < 1$ provided x is sufficiently close to S. Equation (2.31) tells us that if $\lambda < 0$, then $G'(0.5) > 1$; while if $\lambda > 1.266$, then $G'(0.5) < -1$. In both cases divergence would follow. We therefore conclude that convergence requires λ to lie somewhere in the range

$$0 < \lambda < 1.266 \tag{2.32}$$

Any value of λ within this range should lead to convergence (although the use of a value near 0 or near 1.266 – the ends of the range of convergence – would be risky, because these limits were computed using $x = 0.5$, which is only a first approximation to S). The best value of λ is the one which makes $G'(1.5) = 0$. This value is $1.58^{-1} = 0.633$. Use of this value should lead to faster convergence than any other value will*, and in particular faster than the value $\lambda = 1$ which leads to the recovery of the simple formula, (2.21).

---

* Subject, again, to the limitation that it was obtained using the approximation $S = 0.5$.

Table 2.4 Effect of λ on the rate of convergence of (2.29) when used to solve (2.14).

n	|	|	|	λ	|	|
	0.5	0.6	0.60625	0.633	0.7	1
1	0.5	0.5	0.5	0.5	0.5	0.5
2	0.58080	0.59696	0.59797	0.60229	0.61312	0.66160
3	0.59658	0.59997	0.60000	0.59990	0.59793	0.55888
4	0.59940	0.60000	0.60000	0.60000	0.60032	0.62617
error in x_4(%)	-1×10^{-1}	-6×10^{-5}	0	7×10^{-4}	5×10^{-2}	4

Table 2.4 illustrates the rates of convergence with various values of λ, and the errors in the respective estimates of S after three iterations. It is seen that the theoretical value $\lambda = 0.625$ does lead to the fastest convergence: x_4 is, in fact, exact within the 12-digit precision used for the calculations. The use of $\lambda = 0.633$, determined from (2.31), is also very efficient. In contrast, the use of $\lambda = 1$ (which corresponds to undamped, simple iteration) can be shown to require 23 iterations to achieve the same accuracy as that which $\lambda = 0.633$ yields in three.

To generalize this discussion, we note from (2.29) that

$$G'(x_n) = \lambda F'(x_n) + 1 - \lambda$$

To force $G'(x_n)$ to be zero, we choose λ to be

$$\lambda = \frac{1}{1 - F'(x_n)}$$

This implies that λ should be recalculated for each iteration. If this is done, then the process will converge in the minimum number of iterations, but extra work (the calculation of λ) is performed each time. Alternatively, λ can be calculated from x_1 and that value used at each iteration. More iterations will be required, but without the additional work per iteration. Finally (and this is perhaps only feasible if the calculations are being performed on a hand-held calculator), λ can be recomputed every few iterations as a compromise between reducing the number of iterations and increasing the work per iteration.

We found early in this chapter that the iteration process defined by (2.16) did not permit the solution of (2.14) near $x = 4$ to be obtained. However, we were able to find that solution using (2.17). Can we also do so using (2.29)? From (2.30) we obtain

$$G'(4) = 5.186\lambda + 1$$

from which the optimum value of λ is estimated to be -0.193. Students should verify that the use of this value of λ is, indeed, found to lead to a convergent process, with an error in S of just over 0.01% being obtained in three iterations.

2.8 Newton–Raphson method

In deriving (2.21) the proviso was made that $F'(S)$ should not be zero. Clearly this is necessary, no matter how small e_n is, if the quantity $e_n F'(S)$ is to dominate over the remaining terms in the series. But what if $F'(S)$ *is* zero? This may happen, in a particular problem. In such a case

$$e_{n+1} = \frac{e_n^2}{2!} F''(S) + \frac{e_n^3}{3!} F'''(S) + \cdots$$

and provided now that $F''(x) \neq 0$, and that e_n is sufficiently small,

$$e_{n+1} \approx \frac{e_n^2}{2} F''(S) \qquad (2.33)$$

Convergence requires that

$$|e_{n+1}| < |e_n|$$

Using (2.33) which, it must be remembered, is only an approximation, it is therefore approximately true to say that

$$\tfrac{1}{2}|e_n^2 F''(S)| < |e_n|$$

and hence that

$$|e_n| < \frac{2}{|F''(S)|} \qquad (2.34)$$

In particular – and still assuming that (2.33) is true – the first estimate x_0 must, approximately, satisfy

$$|e_0| < \frac{2}{|F''(S)|} \qquad (2.35)$$

Thus, if the iteration formula

$$x_{n+1} = F(x_n) \qquad (2.36)$$

is such that $F'(S) = 0$, and if the initial estimate x satisfies (2.35), then the process will converge and, moreover, will be a *second-order* process: by (2.33)

$$e_{n+1} \propto e_n^2 \qquad (2.37)$$

This is a highly desirable situation, for the errors will now diminish much more rapidly than they do in a first-order process. It is therefore worth trying to find a function $F(x)$ – that is, a rearrangement of the original equation – with the property $F'(S) = 0$.

Since S is a solution of both

$$f(x) = 0 \quad \text{and} \quad x = F(x)$$

then it will also satisfy

$$x = F(x) + H(x)\,f(x)$$

where $H(x)$ is any function of x except one for which $H(S) = \infty$. Therefore

$$F(x) = x - H(x)\,f(x) \tag{2.38}$$

and we will try to choose $H(x)$ to ensure that $F'(S) = 0$.

Now,

$$F'(x) = 1 - H(x)\,f'(x) - H'(x)\,f(x)$$

Hence

$$F'(S) = 1 - H(S)\,f'(S)$$

since $f'(S) = 0$. We are trying to make $F'(S)$ equal to zero. This will happen if $H(S) = 1/f'(S)$. One way to achieve this is to choose

$$H(x) = 1/f'(x) \tag{2.39}$$

Then, as $x \to S$, $H(x) \to 1/f'(S)$. Substitution of (2.39) into (2.38) yields

$$F(x) = x - \frac{f(x)}{f'(x)} \tag{2.40}$$

and the iteration formula (2.36) becomes

$$x_{n+1} = x_n - \frac{f(x_n)}{f'(x_n)} \tag{2.41}$$

This iteration formula is known as the Newton–Raphson method, or simply as Newton's method.

We can readily verify from (2.40) that $F'(S) = 0$:

$$F'(x) = 1 - \frac{\{f'(x)\}^2 - f(x)\,f''(x)}{\{f'(x)\}^2}$$

$$= \frac{f(x)\,f''(x)}{\{f'(x)\}^2} \tag{2.42}$$

Since $f(S) = 0$, it follows that $F'(S) = 0$ *no matter what form the function $f(x)$ takes*.

Hence, Newton's method is at least second order. We should check whether Newton's method is, perhaps, of the third (or even higher) order: is $F''(S) = 0$, by any chance? From (2.42) we find that

NEWTON–RAPHSON METHOD

$$F''(S) = \frac{f''(S)}{f'(S)} \tag{2.43}$$

In general, $f''(S)$ will not be zero, and therefore $F''(S)$ will not, in general, be zero. Thus Newton's method is exactly second order, and not higher.

There is one situation in which $F'(S)$ will not be zero: viz. when $f'(S)$ is zero. In this case, since $f(S)$ is also zero, (2.40) is indeterminate. We will consider this possibility in the next section.

It must again be emphasized that the condition on the initial estimate for convergence of Newton's method is only approximate, since it was derived from the truncation of a Taylor series. In some problems it may not be sufficient, in others it may not be necessary. All that can be said with assurance is that Newton's method will converge for any equation, provided the first estimate of the solution is good enough – but precisely what is meant by 'good enough' cannot be stated in general terms.

A graphical interpretation of Newton's method is shown in Figure 2.9. The value of x at the point A is x_n, an estimate of the solution S. Hence the distance AB is $f(x_n)$. The line BC is drawn tangent to the curve $y = f(x)$ at B, and hence its slope is $f'(x_n)$. In other words,

$$\text{AB/CA} = f'(x_n)$$

and thus

$$\text{CA} = \frac{\text{AB}}{f'(x_n)} = \frac{f(x_n)}{f'(x_n)}$$

Therefore, the value of x at C is

$$x_n - \frac{f(x_n)}{f'(x_n)}$$

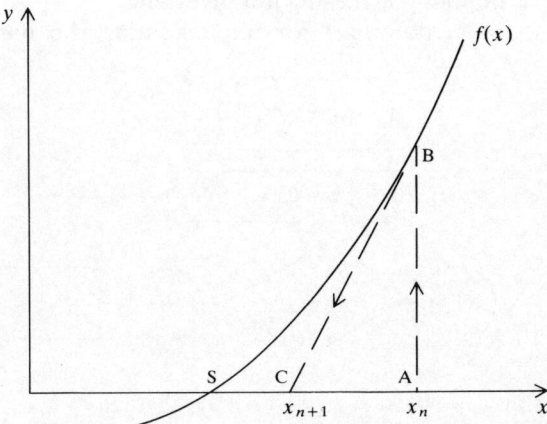

Figure 2.9 Newton's method.

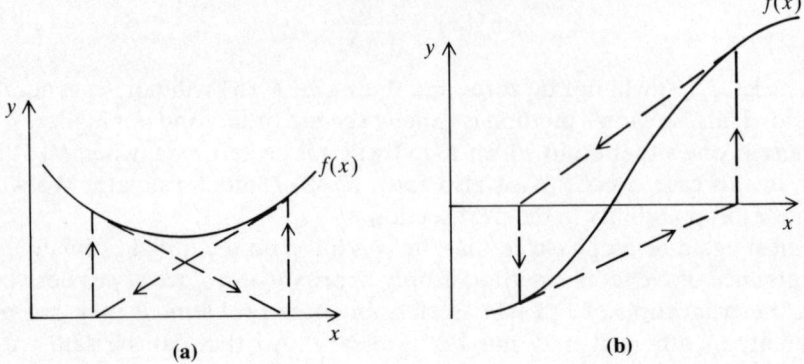

Figure 2.10 Infinite looping with Newton's method.

which by (2.41) is the value of x_{n+1}. Newton's method is seen to be equivalent to locating an improved estimate at the point C by constructing the perpendicular AB and the tangent BC. Repetition of this construction will generate a sequence of points along the x-axis approaching the solution at $x = S$.

A graph is also convenient to illustrate some situations in which Newton's method will fail. Figure 2.10 shows situations in which the iterations will continue indefinitely. In Figure 2.10a there is no real solution of $f(x) = 0$ in the vicinity and, of course, all methods will fail to find a root there. In Figure 2.10b there is a solution, but the successive estimates oscillate indefinitely without approaching it. Unless the curve $y = f(x)$ happens to be antisymmetrical about the solution, this problem can be overcome by locating the first estimate more accurately. This is another situation in which an iterative process is neither converging nor diverging.

Figure 2.11 shows a flowchart for implementing the basic Newton's

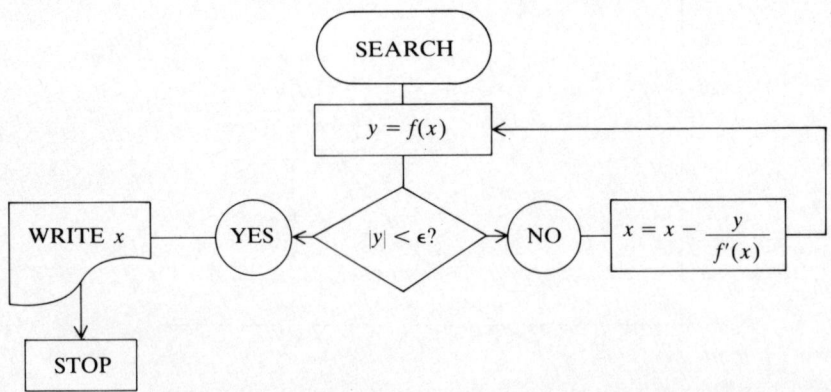

Figure 2.11 A flowchart for the basic Newton's method.

method. The iterations are terminated when $f(x)$ is less than a specified small number. Until then, new values of x are computed according to (2.41).

This flowchart illustrates the essential simplicity of Newton's method. However, there are several modifications which should be made to it.

Consider, first, the test for terminating the iterations. Ideally, we would like to find the value of x which makes $f(x)$ exactly zero. However, we cannot do this. Equation (2.33) shows that if x_0 is different from S, then x_n will also be different from S for all values of n. The error will diminish, but will never be exactly zero. Moreover, since we are forced, with a computer or with any other calculating device, to use only a finite number of significant figures, we cannot in general expect to find the *exact* solution by any process. We must be satisfied with a value of x which makes $f(x)$ 'almost' zero, i.e. within the range $\pm \varepsilon$.

However, even this may not be a satisfactory test. For without knowing the sensitivity of $f(x)$ to a change in x, we do not know how small to choose ε. If $f'(x)$ is small, then larger changes in x will cause only small changes in $f(x)$, and ε will need to be very small. On the other hand, if $f'(x)$ is large, then a small value of ε will cause the iterations to be continued longer than necessary. A similar difficulty was encountered with simple iteration in the previous section.

This difficulty is overcome by noting that $f(x)/f'(x)$, which is the change in x from one iteration to the next, is also a measure of the error remaining in x. For

$$\begin{aligned}\frac{f(x_n)}{f'(x_n)} &= x_n - x_{n+1} \\ &= (S + e_n) - (S + e_{n+1}) \\ &= e_n - e_{n+1} \\ &\approx e_n(1 - Ke_n)\end{aligned}$$

by (2.33), where $K = \tfrac{1}{2}F''(S)$. Thus

$$e_n = \frac{f(x_n)/f'(x_n)}{1 - Ke_n}$$

As $n \to \infty$, Ke_n becomes small compared with unity; therefore

$$e_n \to \frac{f(x_n)}{f'(x_n)}$$

and hence

$$x_n - x_{n+1} \to e_n \qquad (2.44)$$

Normally we are more interested in the nearness of x to S than in the smallness of $f(x)$. Equation (2.44) shows that $x_n - x_{n+1}$ measures this. But how small should e_n be? Clearly this depends on the magnitude of S itself: we should again use a *relative* test. We would like to know that the 'solution' is accurate to, say, 0.1%. Thus a criterion for terminating the iterations could be

$$|x_{n+1} - x_n| < |x_{n+1}/1000|$$

or, in general

$$|x_{n+1} - x_n| < \varepsilon |x_{n+1}| \tag{2.45}$$

where ε is a small number which we specify.

We should also guard against the possibility that the process will never converge, for example if either of the situations shown in Figure 2.10 should arise. It is necessary to count the number of iterations performed, and to stop when some specified upper limit has been reached. A warning message should be issued by the program, so that we will know something has gone wrong and can try to find out what. Under normal conditions Newton's method should converge to an acceptable accuracy in 10 or 15 iterations; an upper limit of, say, 20 would therefore be appropriate. Obviously, there will be occasions on which this limit will need to be increased.

Another aspect of our present algorithm which should be considered concerns the calculation of $f'(x)$. If the function $f(x)$ is fairly simple, then an analytical expression for $f'(x)$ can be readily derived and programmed and the value for the derivative computed from it. However, if $f(x)$ is complicated, then although $f'(x)$ can still be found, the evaluation may prove to be time-consuming. An acceptable alternative is to approximate $f'(x)$ by

$$f'(x) = \frac{f(x + \delta x) - f(x)}{\delta x}$$

where δx is a small quantity. For a reason similar to that discussed above in relation to terminating the iterations, δx should be specified as some fraction – say 0.01 – of the current value of x. Our iteration equation thus becomes

$$x_{n+1} = x_n - \frac{0.01 x_n f(x_n)}{f(1.01 x_n) - f(x_n)} \tag{2.46}$$

The development of a flowchart and the resulting computer program, incorporating (2.45) and (2.46), are left as exercises for the student.

As an example of the rates of convergence of simple iteration and Newton's method, consider the equation

$$f(x) \equiv x^2 - 3x + 2 = 0$$

Writing this as

$$x = (3x - 2)^{1/2} = F(x)$$

and noting that one solution occurs at $x = 2$, we see that

$$F'(2) = 0.75$$

EXTENDED NEWTON'S METHOD

Table 2.5 Comparison of the solution of $x^2 - 3x + 2 = 0$ by simple iteration and by Newton's method.

	Simple iteration			Newton's method		
n	x_n	e_n	$0.75 e_{n-1}$	x_n	e_n	e_{n-1}^2
0	2.1	0.1		2.1	0.1	
1	2.0736	0.0736	0.075	2.008 333	0.008 333	0.01
2	2.0545	0.0545	0.0552	2.000 068	0.000 068	0.000 069
3	2.0405	0.0405	0.0409	2.000 000		
4	2.0301	0.0301	0.0304			
⋮	⋮	⋮	⋮			
9	2.0070	0.0070				
10	2.0052	0.0052	0.0053			

Hence, for simple iteration we expect, from (2.21), that

$$e_{n+1} = 0.75 e_n \tag{2.47}$$

For Newton's method, using (2.43),

$$F''(x) = \frac{f''(x)}{f'(x)} = \frac{2}{2(2) - 3} = 2$$

Hence (2.33) yields

$$e_{n+1} = e_n^2 \tag{2.48}$$

Note that this simple result applies only to the particular example being considered at the moment.

Table 2.5 presents the results for several iterations of each method, including a comparison between the actual errors at each iteration and the respective predictions from (2.47) and (2.48). Since the predicted behaviour of the errors is strictly true only in the limit when $n \to \infty$ (i.e. when the neglected terms in the Taylor series expansions really are negligible), the results are seen to be in good agreement with the theory.

2.9 Extended Newton's method

We have seen that if $f'(S) = 0$, Newton's method may fail. Indeed, if at any stage of the iterations $f'(x_n)$ becomes very small, difficulties may arise; $x_{n+1} - x_n$ can become large, and x_{n+1} may then be a worse estimate, rather than a better one, of the solution being sought. There are a number of situations in which $f'(x_n)$ can become small, and these are illustrated in Figure 2.12.

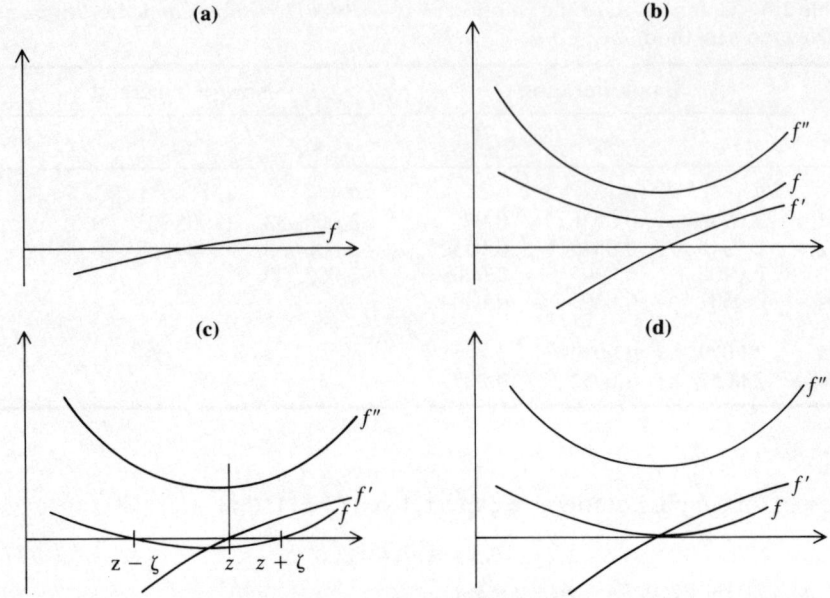

Figure 2.12 Situations in which $f'(x)$ is small.

In Figure 2.12a there is no real difficulty, because although $f'(x_n)$ may be small, it does not become zero; and in any case $f(x_n)$ also becomes small, and vanishes at $x = S$. So Newton's method should still be well-behaved.

In Figures 2.12b–d, however, the possibility exists of $f'(x_n)$ becoming vanishingly small while $f(x_n)$ is also small or even zero. These three cases require special treatment.

It can be seen that they will arise if, respectively, $f(x)$ has no zeros, two distinct zeros or two equal zeros in the neighbourhood of the point where

$$f'(x) = 0 \tag{2.49}$$

The first step is to determine just where this occurs, by solving the subsidiary problem (2.49). Call the solution Z. We can then distinguish between the three situations by examining $f''(Z)$ since, as can be seen from Figure 2.12, there will be no solutions, two different solutions or two equal solutions of $f(x) = 0$ near $x = Z$, depending on whether

$$f(Z) \cdot f''(Z) > 0 \tag{2.50}$$

$$f(Z) \cdot f''(Z) < 0 \tag{2.51}$$

or

$$f(Z) = 0 \tag{2.52}$$

respectively. (It should be noted that identical conditions will apply when the curves of Figures 2.12b–d are reflected in the x-axis. Students should verify this by drawing the corresponding sketches.)

If (2.50) is found to be true, then the search for a solution near Z will be abandoned. If (2.52) is true, then the search is over – a double root has been located. If (2.51) applies, then estimates of the two solutions near Z can be found. Let one of these be at $S = Z - \zeta$. Then

$$f(S) = 0$$

and therefore

$$f(Z - \zeta) = 0 = f(Z) - \zeta f'(Z) + \frac{\zeta^2}{2} f''(Z) - \cdots$$

Noting that $f'(Z) = 0$, and assuming that terms in the third and higher powers of ζ can be neglected, then

$$0 = f(Z) + \frac{\zeta^2}{2} f''(Z)$$

and, hence,

$$\zeta = \pm\{-2f(Z)/f''(Z)\}^{1/2}.$$

Estimates of the two solutions near $x = Z$ are therefore

$$x = Z \pm \{-2f(Z)/f''(Z)\}^{1/2}. \tag{2.53}$$

By truncating the series expansion of $f(Z - \zeta)$ after the third term we are, in effect, assuming that $f(x)$ is locally quadratic. Hence the two solutions given by (2.53) are equidistant from Z. These solutions may be accepted as they are or, preferably, may be subjected to further refinement by separate applications of Newton's method. Their accuracy will normally be such that further trouble from smallness of $f(x)/f'(x)$ will not be encountered.

Difficulties may also arise if, in the neighbourhood of some point Z, there are multiple roots of the third or higher order. It is not generally worth modifying the program to deal with these situations although, of course, this can be done.

2.10 Other iterative methods

Newton's method requires the calculation, each cycle, of two function values; either $f(x_n)$ and $f'(x_n)$, or $f(x_n)$ and $f(x_n + \delta x)$. Also, we have seen that if $f'(x_n)$ is small, then complications can arise.

In an attempt to reduce the amount of work involved in the solution process, and also to avoid the trouble associated with small derivatives, other iterative techniques have been developed. These are particularly useful when calculators, rather than computers, are being used.

THE SOLUTION OF EQUATIONS

The *secant method*, illustrated in Figure 2.13, relies on locating the points x_1 and x_2 on either side of the solution. Then x_3 is chosen as the point which lies at the intersection of the x-axis and the straight line – the secant – joining the point $(x_1, f(x_1))$ to the point $(x_2, f(x_2))$. As with interval halving, the sign of $f(x_3)$ is used to determine which of the two intervals (x_1, x_3) or (x_3, x_2) now contains the solution, thus ensuring convergence.

From the geometry of Figure 2.13 it is straightforward to show, for two estimates x_{n-1} and x_n such that $x_{n-1} < S < x_n$, that

$$x_{n+1} = x_n - f(x_n) \frac{x_n - x_{n-1}}{f(x_n) - f(x_{n-1})} \tag{2.54}$$

Note that the signs of $f(x_{n-1})$ and $f(x_n)$, which are different, must be taken into account.

Comparison of (2.54) with (2.41) reveals that the former approximates $f'(x_n)$ by

$$\frac{f(x_n) - f(x_{n-1})}{x_n - x_{n-1}}$$

It can be shown that, as a result, the error in x_n is now given by

$$e_{n+1} \propto e_n e_{n-1} \tag{2.55}$$

compared with (2.37) for Newton's method. The order of the method is somewhere between first and second: that is, it converges in fewer steps than simple iteration (which is first order) but in more steps than Newton's method (which is second order). Despite requiring more steps than Newton's method, the total computation time may well be less, since only one function evaluation is required per step, compared with two for Newton's method.

The method known as *regula falsi*, or the method of false position, is similar to the secant method. It does not require the location of points on

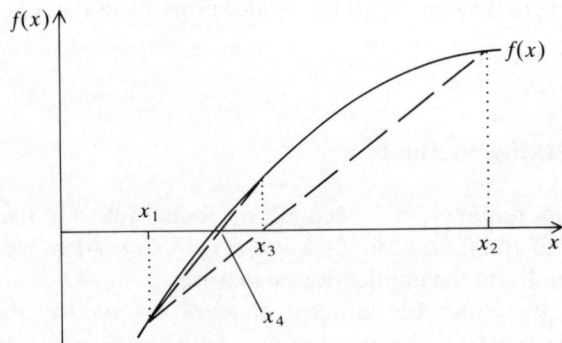

Figure 2.13 The secant method.

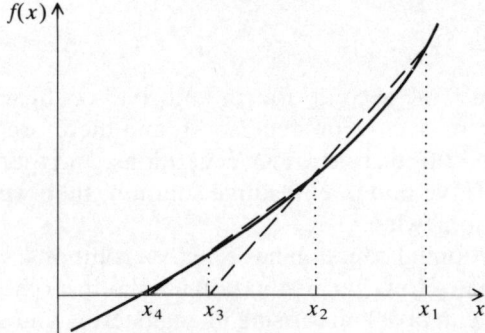

Figure 2.14 The method of false position.

either side of the solution, but merely any two estimates of the solution. It is illustrated in Figure 2.14. Values of $f(x)$ at these two estimates are *extrapolated* to zero, rather than *interpolated*, as in the secant method. Either the two most recent estimates of x or (in some versions) the two estimates which yield the smallest values of $f(x)$ are used to locate the next estimate. Convergence is no longer guaranteed and, although some additional work is involved in the secant method (in determining which of the two intervals contains the solution), it is superior to *regula falsi*. Interval halving is superior to both methods.

2.11 Polynomial equations

Polynomial equations, i.e. equations of the form

$$P(x) = \sum_{j=0}^{n} a_j x^j$$
$$= a_0 + a_1 x + a_2 x^2 + \cdots + a_n x^n = 0 \qquad (2.56)$$

can, of course, be solved by any of the foregoing methods. However, special techniques also exist which take advantage of the nature of such equations. They depend on factorizing the polynomial $P(x)$, and in particular on finding either linear or quadratic factors from which the roots can be immediately obtained.

The approximate location of the roots can sometimes be facilitated by *Descartes' Rule of Signs*:

The number of *positive roots* of a polynomial equation is equal to the number of *sign changes* in the coefficients of the equation, or is less than that by an even number; the number of *negative* roots is equal to the number of *sign repetitions*, or is less than that by an even number.

Thus the equation

$$x^5 - 3x^4 - 13x^3 + 7x^2 + 4x - 12 = 0$$

has three sign changes (the second, fourth and sixth coefficients being of opposite sign to their respective predecessors), and therefore has three or one positive roots; and there are two sign repetitions, and therefore two or zero negative roots. If we find one negative solution, then we know there must be another one somewhere.

An nth degree polynomial equation has exactly n solutions, some or all of which may not be real. However, if the polynomial has real coefficients, which is generally true in problems arising in engineering and science, then any complex solutions must occur in pairs which are complex conjugates:

$$a + ib \quad \text{and} \quad a - ib$$

since it is only under these conditions that the factors

$$(x - a - ib) \quad \text{and} \quad (x - a + ib)$$

will yield only real terms when multiplied together. Thus, we can find any complex roots of a polynomial if we can find the corresponding quadratic factor of $P(x)$. Bairstow's method, described in the next section, achieves this.

Another aid to the location of the solutions is provided by the set of rules sometimes known as *Newton's Relations*. If the n zeros of $P(x)$ are $S_1, S_2, \ldots, S_n$, then it can be shown that

$$\frac{a_{n-1}}{a_n} = -\sum_{j=1}^{n} S_j \tag{2.57}$$
$$= -(\text{the sum of the roots})$$

$$\frac{a_{n-2}}{a_n} = +\sum_{j=1}^{n-1} \sum_{k=j+1}^{n} S_j S_k \tag{2.58}$$
$$= +(\text{the sum of the products of the roots taken two at a time})$$

$$\frac{a_{n-3}}{a_n} = -\sum_{j=1}^{n-2} \sum_{k=j+1}^{n-1} \sum_{l=k+1}^{n} S_j S_k S_l \tag{2.59}$$
$$= -(\text{the sum of the products of the roots taken three at a time})$$

$\vdots$

$$\frac{a_1}{a_n} = (-1)^{n-1} \sum_{j=1}^{2} \sum_{k=j+1}^{3} \cdots \sum_{m=l+1}^{n} S_j S_k \cdots S_l S_m \tag{2.60}$$
$$= (-1)^{n-1} [\text{the sum of the products of the roots taken } (n-1) \text{ at a time}]$$

$$\frac{a_0}{a_n} = (-1)^n \prod_{j=1}^{n} S_j \tag{2.61}$$
$$= (-1)^n (\text{the product of all the roots})$$

To see this, consider the case $n = 4$:

$$\begin{aligned} P(x) &= (x - S_1)(x - S_2)(x - S_3)(x - S_4) \\ &= x^4 - (S_1 + S_2 + S_3 + S_4)x^3 \\ &\quad + (S_1S_2 + S_1S_3 + S_1S_4 + S_2S_3 + S_2S_4 + S_3S_4)x^2 \\ &\quad - (S_1S_2S_3 + S_1S_2S_4 + S_1S_3S_4 + S_2S_3S_4)x \\ &\quad + S_1S_2S_3S_4 \end{aligned} \quad (2.62)$$

By associating the coefficients in (2.62) with those in (2.56), Newton's Relations follow.

They are particularly useful in two situations. First, if one of the solutions, S_{max} (say), is much larger (in absolute value) than all the others, then it will dominate the right-hand side of (2.57). Thus, a first estimate of this solution is

$$S_{max} \approx -a_{n-1}/a_n \qquad (2.63)$$

Secondly, suppose that one of the solutions, S_{min} (say), is much smaller (in absolute value) than all the others. Now each of the terms on the right-hand side of (2.60) are products of all but one of the roots, and S_{min} will appear in all but one of these terms. [Inspection of (2.62) should make this easier to see.] Thus, if each side of (2.60) is divided by the respective side of (2.61) the result is

$$\frac{a_1}{a_0} = -\sum_{j=1}^{n} \frac{1}{S_j} \qquad (2.64)$$

Since $(1/S_{min})$ will be larger than all the other terms on the right-hand side of (2.64), it follows that

$$\frac{a_1}{a_0} = -\frac{1}{S_{min}}$$

and therefore that

$$S_{min} \approx -(a_0/a_1) \qquad (2.65)$$

For example, if

$$P(x) = x^4 - 120x^3 + 2109x^2 - 10\,990x + 9000 \qquad (2.66)$$

(2.63) and (2.65) yield

$$S_{max} \approx -(-120)/(1) = 120$$

and

$$S_{min} \approx -(9000)/(-10\,990) = 0.82$$

The zeros of (2.66) are actually 100, 10, 9 and 1, so the estimates S_{max} and S_{min} are quite good. If the solutions happen to be more closely spaced, then Newton's Relations cannot be expected to yield good first estimates – because of the approximations made – but they may still be better than mere guesswork.

The procedure for locating factors of a polynomial $P(x)$ is as follows: we try to find a term of the general form $(x - a)$ which gives a zero remainder when divided into $P(x)$. Since at best we only know the factor approximately in the first instance, we need an iterative procedure which will allow us to improve the approximation. Moreover, we need a procedure which can readily be automated. The method known as *synthetic division* satisfies these requirements.

The method is best understood by first considering an example. We will divide (2.66) by $(x - 0.82)$ which, as we have just seen, should be an approximation to a factor of the polynomial.

$$
\begin{array}{r}
x^3 - 119.18x^2 + 2011.27x - 9340.76 \\[4pt]
x - 0.82 \overline{\smash{)} x^4 - 120\ x^3 + 2109\ x^2 - 10\,990\ x + 9000} \\
x^4 - 0.82x^3 \\ \hline
-119.18x^3 + 2109\ x^2 \\
-119.18x^3 + 97.73x^2 \\ \hline
2011.27x^2 - 10\,990\ x \\
2011.27x^2 - 1649.24x \\ \hline
-9340.76x + 9000 \\
-9340.76x + 7659.42 \\ \hline
1340.58
\end{array}
$$

The result is

$$x^3 - 119.18x^2 + 2011.27x - 9340.76$$

with a remainder of 1340.58, showing that $(x - 0.82)$ is *not* a factor of $P(x)$.

The actual mechanics of the division process can be considerably simplified by omitting non-essential features. For example, in each subtraction the coefficients of the higher power of x are the same; indeed, we choose the quotient to force this to be true. Thus, we need not bother to write down these higher-powered terms. Furthermore, we do not really need to write the xs down at all. We know that the powers of x in each column are the same, decreasing from left to right, so we only need to retain the coefficients. Finally, if we change the sign of the constant term in the divisor (i.e. if we replace -0.82 by $+0.82$ in the example) and replace the subtraction operations by additions, we will achieve the same end-result. The process can now be written more economically as a modified form of simple division rather than long division:

$$
\begin{array}{r|rrrrr}
0.82 & 1 & -120 & 2109 & -10\,990 & 9000 \\
 & & 0.82 & -97.73 & 1649.24 & -7659.42 \\ \hline
 & 1 & -119.18 & 2011.27 & -9340.76 & 1340.58
\end{array}
$$

A careful comparison of these two forms of the process will reveal that it can be generalized in the following way:

POLYNOMIAL EQUATIONS

$$\begin{array}{c|ccccccc}
x_0 & a_n & a_{n-1} & a_{n-2} & \cdots & a_1 & a_0 \\
 & & b_n x_0 & b_{n-1} x_0 & \cdots & b_2 x_0 & b_1 x_0 \\
\hline
 & b_n & b_{n-1} & b_{n-2} & \cdots & b_1 & b_0
\end{array}$$

where the quantities b_j are calculated from the recursion formula

$$b_n = a_n$$
$$b_j = a_j + b_{j+1} x_0 \qquad j = n-1, n-2, \ldots, 0 \qquad (2.67)$$

The quotient is a polynomial of degree $n - 1$:

$$Q(x) = \sum_{j=1}^{n} b_j x^{j-1}$$

and there is a remainder of b_0. Thus,

$$\frac{P(x)}{x - x_0} = Q(x) + \frac{b_0}{x - x_0} \qquad (2.68)$$

or

$$P(x) = (x - x_0)Q(x) + b_0$$

When $x = x_0$, $P(x_0) = b_0$; that is, b_0 is the value of $P(x)$ at $x = x_0$. If $(x - x_0)$ is a factor of $P(x)$, then b_0 will be zero, i.e. $P(x_0) = 0$.

The algorithm defined by (2.67) is a simple, easily programmed procedure for evaluating a polynomial for a specified value of x. If the result, b_0, is not zero, then the value of x_0 must be modified. To see how this may be done, consider the division of $Q(x)$ by $(x - x_0)$. The result will be a polynomial, $R(x)$ (say), of degree $n - 2$, and a remainder, c_1:

$$\frac{Q(x)}{x - x_0} = R(x) + \frac{c_1}{x - x_0}$$

or

$$Q(x) = (x - x_0)R(x) + c_1$$

Therefore, from (2.68),

$$\frac{P(x)}{x - x_0} = (x - x_0)R(x) + c_1 + \frac{b_0}{x - x_0}$$

or

$$P(x) = (x - x_0)^2 R(x) + (x - x_0)c_1 + b_0$$

Therefore,

$$\frac{d}{dx} P(x) = P'(x) = 2(x - x_0)R(x) + (x - x_0)^2 R'(x) + c_1$$

and when $x = x_0$ we find that $P'(x_0) = c_1$. In other words, c_1 is the value of the derivative $P'(x)$ at $x = x_0$.

The coefficients, c_j (say), of $R(x)$ are obtained from those of $Q(x)$ by synthetic division again. By analogy with (2.67),

$$c_n = b_n$$
$$c_j = b_j + c_{j+1}x_0, \quad j = n-1, n-2, \ldots, 1 \qquad (2.69)$$

Note that there are $(n+1)$ coefficients $(a_n, a_{n-1}, \ldots, a_0)$ of $P(x)$, there are n coefficients $(b_n, b_{n-1}, \ldots, b_1)$ of $Q(x)$ together with the remainder b_0, and there are $(n-1)$ coefficients $(c_n, c_{n-1}, \ldots, c_2)$ of $R(x)$ together with the remainder c_1.

While we need to know all of the coefficients b_j of $Q(x)$, so that $R(x_0)$ can be evaluated, we do not need to know all of the coefficients of $R(x)$. The only one of interest is the last value, c_1. Thus, bearing in mind that a computer program is to be written to implement this algorithm, the successive values of c_j can be overwritten during the computation and (2.69) becomes

$$c = b_n$$
$$c = b_j + cx_0, \quad j = n-1, n-2, \ldots, 1 \qquad (2.70)$$

The two values $P(x_0) = b_0$ and $P'(x_0) = c$ enable Newton's method to be used for the solution of $P(x) = 0$:

$$x_{k+1} = x_k - \frac{P(x_k)}{P'(x_k)} = x_k - \frac{b_0(x_k)}{c(x_k)} \qquad (2.71)$$

where the dependence of b_0 and c on x_k has been emphasized.

To illustrate this, let us continue the example started before:

```
0.82  | 1  -120      2109      -10990       9000
      |      0.82   -97.73    1649.24    -7659.42
      |_____
0.82  | 1  -119.18  2011.27   -9340.76    1340.58 = P(0.82)
      |      0.82    97.06    1569.66
      |_____
        1  -118.36  1914.22   -7771.10 = P'(0.82)
```

The fact that $P(0.82) = 1340.58$ and $P'(0.82) = -7771.10$ can be readily verified by substitution.

Equation (2.71) now tells us that

$$x_1 = 0.82 - (1340.58)/(-7771.10) = 0.9925$$

Let us perform one more step of the process:

```
0.9925 | 1  -120       2109       -10990         9000
       |      0.9925  -118.1149  1975.9534   -8946.4412
       |_____
0.9925 | 1  -119.0075  1990.8851  -9014.0466   53.5588 = P(0.9925)
       |      0.9925   -117.1299  1859.7020
       |_____
         1  -118.0150  1873.7552  -7154.3446 = P'(0.9925)
```

Thus
$$x_2 = 0.9925 + 53.5588/7154.3446 = 1.0000$$

The process has converged to the solution $x = 1$ (to five significant figures, at any rate) in two iterations.

This combination of synthetic division and Newton's method is very suitable for use with a calculator, and is also easily implemented on a digital computer.

Figure 2.15 shows the flowchart.

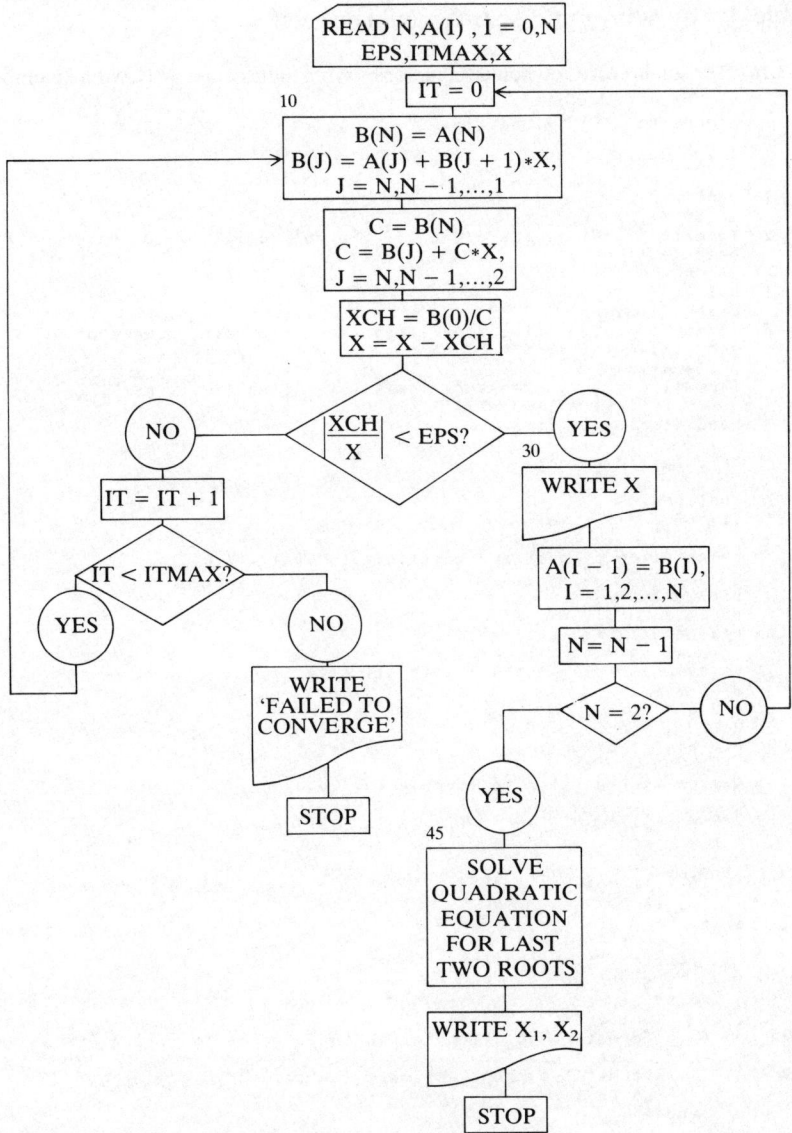

Figure 2.15 A flowchart for the solution of a polynomial equation by synthetic division and Newton's method.

The input to the program consists of

    'x' – the initial estimate of the solution
    'n' – the degree of the polynomial $P(x)$
    'a(i)' – the $(N + 1)$ coefficients of $P(x)$
    'eps' – the convergence criterion and
  'itmax' – the maximum number of iterations permitted.

A program based on this flowchart is shown in Figure 2.16, together with an example of its use. It is assumed that initial estimates of the solutions are available. If not, subroutine 'search' could be used.

Figure 2.16   The solution of a polynomial equation by synthetic division and Newton's method.

```
c
          dimension a(0:20), b(0:19)
c
c         data input
c
      1   continue
          write (*,2)
      2   format (' what is the degree of the polynomial? ( <0 to stop)')
          read (*,*) n
          if (n .le. 0) stop
          np1 = n + 1
          write (*,3) np1
      3   format (' enter ',i3,' coefficients in ascending powers of x')
          read (*,*) (a(i), i = 0, n)
          write (*,5)
      5   format(' enter convergence limit, iteration limit and ',
         +       'first estimate')
          read (*,*) eps, itmax, x
c
c         calculations start
c
      7   continue
          it = 0
c
c         calculation of quotient coefficients b(n) ... b(0)
c
     10   b(n) = a(n)
          do 15 j = n-1, 0, -1
     15   b(j) = a(j) + b(j+1)*x
c
c         calculation of all the c's to get c(1)
c
          c = b(n)
          do 20 j = n-1, 1, -1
     20   c = b(j) + c*x
c
c         Newton-Raphson step to improve the value of x
c
          xch = b(0)/c
          x = x - xch
c
          if( abs (xch/x) .lt. eps) then
             go to 30
          else
             it = it + 1
             if (it .lt. itmax) then
                go to 10
             else
                write (*,25) itmax
     25         format(/' solution failed to converge in ',i3,' iterations'
                write (*,26) x
     26         format(' current estimate is ',g10.3/)
                go to 1
             endif
          endif
c
c         c o n v e r g e d
```

```
c
   30  write (*,35) it, x
   35  format(' a solution in ',i3,' iterations is ',f7.4)
       write(*,36) b(0)
   36  format(' for which   f(x) = ',e11.4)
c
c      generate reduced polynomial
c
       do 40 i = 1, n
   40  a(i-1) = b(i)
       n = n - 1
       write (*,42)
   42  format(/' the coefficients of the reduced polynomial are:'/)
       write (*,43) (a(i), i = 0, n)
   43  format( 10f10.4 )
       if (n .eq. 2) go to 45
       write (*,44)
   44  format(' next solution:')
       go to 7
c
c      last two roots - solve the quadratic equation directly
c
   45  disc = a(1)**2 - 4.*a(0)*a(2)
       if ( disc .ge. 0.) go to 47
       write (*,46)
   46  format(' sorry - last two roots are complex')
       go to 100
   47  x1 = (-a(1) + sqrt(disc)) / ( 2.*a(2))
       x2 = (-a(1) - sqrt(disc)) / ( 2.*a(2))
       write (*,50) x1, x2
   50  format(/' the last two solutions are ',f7.4,' and ',f7.4)
c
c      return to beginning for another equation
c
  100  go to 1
       end
```

The dialogue with the program in Figure 2.16 might appear as follows:

```
what is the degree of the polynomial? ( <0  to stop)
        3
enter    4 coefficients in ascending powers of x
  -7.0     -7.0       0.0      1.0
enter convergence limit, iteration limit and first estimate
  0.00001       20        1.0
a solution in   6 iterations is -1.6920
for which   f(x) = -0.1431E-05

the coefficients of the reduced polynomial are:

  -4.1371    -1.6920    1.0000

the last two solutions are  3.0489 and -1.3569
what is the degree of the polynomial? ( <0  to stop)
        4
enter    5 coefficients in ascending powers of x
   6.1      1.0       2.0     -4.0      1.0
enter convergence limit, iteration limit and first estimate
  0.00001       20        1.0
a solution in   3 iterations is  2.9922
for which  f(x) =   0.1149E-03

the coefficients of the reduced polynomial are:

  -2.0386    -1.0155   -1.0078    1.0000
next solution:
a solution in   4 iterations is  2.0143
for which  f(x) =   0.2837E-04

the coefficients of the reduced polynomial are:

   1.0121     1.0066    1.0000
sorry - last two roots are complex
what is the degree of the polynomial? ( <0  to stop)
       -1
```

The program allows for all the solutions of $P(x) = 0$ to be obtained (assuming that they are real). When a solution $x = S$ has been found, the polynomial $Q(x)$ is obtained, which is the result of dividing $P(x)$ by the factor $(x - S)$. Thus, the remaining roots of $P(x) = 0$ are also roots of $Q(x) = 0$. $Q(x)$ is called the *reduced* or *deflated polynomial*, and its zeros can be found by the same program. All that is necessary is to redefine the coefficients 'a(i)' to be the latest values of 'b(i)', noting that there is now one coefficient less, so that N must be reduced by 1. Finally, when $N = 2$ and the polynomial is a quadratic, the two last solutions are obtained directly.

2.12 Bairstow's method

The technique just described has one obvious drawback: it will fail if $P(x)$ has complex zeros. However, as was pointed out above, if there are complex roots and if $P(x)$ has only real coefficients, then these roots must occur in complex conjugate pairs. Thus $P(x)$ will always possess real *quadratic* factors, from which two solutions – real or complex – can be found directly. *Bairstow's* method is a modification of synthetic division which looks for quadratic factors of a polynomial.

It is also an iterative technique: an initial estimate

$$x^2 - s_1 x - t_1$$

is somehow found. This is not as easy as it sounds, nor as easy as finding an estimate of a single root of an equation. Some starting values for s and t are necessary: $s = t = 1$ may be as good as any, if there is no hint from the problem itself. A sequence (s_k, t_k) of pairs of values is then formed as follows.

The given polynomial is more conveniently written now as

$$P(x) = a_0 x^n + a_1 x^{n-1} + \cdots + a_n$$
$$= \sum_{j=0}^{n} a_j x^{n-j}, \quad n > 2 \qquad (2.72)$$

The quantities b_j, $j = 0, 1, \ldots, n$ and c_j, $j = 0, 1, \ldots, n-1$ are then computed from

$$b_j = a_j + s_k b_{j-1} + t_k b_{j-2}$$
$$c_j = b_j + s_k c_{j-1} + t_k c_{j-2}$$

with $b_{-1} = b_{-2} = c_{-1} = c_{-2} = 0$.

The quantities δ and ε are then found from

$$\delta = \frac{b_n c_{n-3} - b_{n-1} c_{n-2}}{c_{n-2}^2 - c_{n-1} c_{n-3}} \qquad \varepsilon = \frac{b_{n-1} c_{n-1} - b_n c_{n-2}}{c_{n-2}^2 - c_{n-1} c_{n-3}}$$

It can be shown* that

$$s_{k+1} = s_k + \delta \quad \text{and} \quad t_{k+1} = t_k + \varepsilon$$

are improved estimates of the coefficients of the quadratic factor $(x^2 - sx - t)$ of the polynomial (2.72).

Convergence is guaranteed if the initial estimates s_1 and t_1 are 'sufficiently' good – but, as is often the case, we do not know what that means quantitatively. After convergence has been obtained, the reduced polynomial is

$$b_0 x^{n-2} + b_1 x^{n-3} + \cdots + b_{n-3} x + b_{n-2}$$

Bairstow's method can now be applied to this polynomial until the reduced polynomial is a quadratic or linear expression which can be solved immediately.

Figure 2.17 shows a Fortran program which finds the coefficients in a quadratic factor of a polynomial. It is left to the student to extend the program to enable it to find the subsequent factors of the reduced polynomial, and also to compute and print the actual roots: the zeros (real or complex) of each quadratic factor.

* See, for instance, Isaacson & Keller (1966).

Figure 2.17 The solution of a polynomial equation by synthetic division and Bairstow's method.

```
c
      dimension a(0:20), b(-2:20), c(-2:19)
c
c     data input
c
    1 continue
      write (*,2)
    2 format (' what is the degree of the polynomial? ( <0 to stop)')
      read (*,*) n
      if (n .le. 0) stop
      np1 = n + 1
      write (*,3) np1
    3 format (' enter ',i3,' coefficients in descending powers of x')
      read (*,*) (a(i), i = 0, n)
      write (*,5)
    5 format(' enter convergence limit, iteration limit and ',
     +       'first estimates')
      read (*,*) eps, itmax, s, t
c
c     calculations start
c
      b(-2) = 0.
      b(-1) = 0.
      c(-2) = 0.
      c(-1) = 0.
    7 continue
      it = 0
    8 do 10 j = 0, n
   10 b(j) = a(j) + s*b(j-1) + t*b(j-2)
      do 20 j = 0, n-1
   20 c(j) = b(j) + s*c(j-1) + t*c(j-2)
c
      bottom_line = c(n-2)*c(n-2) - c(n-1)*c(n-3)
```

(continued)

```
              delta = (b(n)*c(n-3) - b(n-1)*c(n-2)) / bottom_line
              epsilon = (b(n-1)*c(n-1) - b(n)*c(n-2)) / bottom_line
      c
              s = s + delta
              t = t + epsilon
      c
      c       converged ?
      c
              if (abs(delta) .lt. eps .and. abs(epsilon) .lt. eps) then
                  write(*,25) s, t
         25       format(/' ***** converged:  s = ',g10.4,'    t = ',g10.4/)
                  go to 1
              endif
      c
      c
              write(*,30) delta, epsilon, s, t
         30   format(' corrections: ',2g10.2,'   new estimates: ',2g10.4)
      c
              it = it + 1
              if( it .le. itmax ) go to 8
              write (*,40) itmax
         40   format(/' solution failed to converge in ',i3,' iterations')
              write (*,50) s, t
         50   format(' current estimates are ',2g10.2/)
              go to 1
              end
```

Figure 2.17—*continued*

The dialogue with the program of Figure 2.17 might appear as follows:

```
what is the degree of the polynomial? ( <0  to stop)
           3
enter   4 coefficients in descending powers of x
 1.0      0.00E+00    -7.0       -7.0
enter convergence limit, iteration limit and first estimates
 0.10E-04  20   1.0         1.0

corrections:   -0.17       5.3      new estimates: 0.8333      6.333
corrections:    0.35      -0.61     new estimates: 1.184       5.721
corrections:    0.12      -0.41     new estimates: 1.306       5.308
corrections:    0.44E-01  -0.13     new estimates: 1.351       5.177
corrections:    0.60E-02  -0.18E-01 new estimates: 1.357       5.159
corrections:    0.11E-03  -0.35E-03 new estimates: 1.357       5.159

***** converged:   s =   1.357       t =   5.159

what is the degree of the polynomial? ( <0  to stop)
          -1
```

Worked examples

1. Obtain a first estimate of a solution of the following equations.

(a) $x^4 + 2/x^4 = 1000$

For $x > 1$, $x^4 \gg 2/x^4$, therefore neglect the second term on the left-hand side (LHS):

$$x^4 \approx 1000 \quad \text{and} \quad x \approx 5.6$$

(Solution: 5.62341)

(b) $x \sin x = 50$

$|\sin x| \leq 1$; therefore $x \geq 50$. Now $\sin x > 0$ only for $2n\pi \leq x \leq (2n+1)\pi$. Therefore x will be about $(2n + \frac{1}{2})\pi$ and also greater than 50; i.e.

$$x \approx 51.8, 58.1, \text{ etc.}$$

(Solutions: 51.5875, 57.5999, etc.)

(c) $x + x^x = 100$

For $x \gg 1$, $x^x \gg x$; therefore neglect the first term on the LHS. Trial-and-error then shows $x \approx 3\frac{1}{2}$.
(Solution: 3.5813)

2. For each of the equations of Example 1, find an interval Δx within which a solution is located. Then solve the equation by interval halving, obtaining the solution to within 0.1%.

(a) For $x = 5.6$, $x^4 + 2/x^4 = 983.4$
For $x = 5.7$, $x^4 + 2/x^4 = 1055.6$

Therefore x is in the interval (5.6, 5.7).

x_1	x_m	x_r	$f(x_m)$	$\dfrac{x_r - x_1}{2x_m} \times 100$
5.6	5.65	5.7	1019.0480	0.88
5.6	5.625	5.65	1001.1311	0.45
5.6	5.6125	5.625	992.2619	0.22
5.6125	5.6188	5.625	996.7246	0.11
5.6188	5.6219	5.625	998.9260	0.06

Note (1) The maximum possible error in x_m is $\pm(x_r - x_1)/2$.
(2) The calculations are worked only to four significant figures.
(3) The second estimate (i.e. 5.625) happens to be correct within 0.1% – but we had no way of knowing this.
(4) The third to fifth estimates are actually *less* accurate than the second, but we are nevertheless narrowing the range of possible values.

(b) For $x = 51$, $x \sin x = 34.2$
For $x = 52$, $x \sin x = 51.3$

Therefore x is in the interval (51, 52).

x_1	x_m	x_r	$f(x_m)$	$\dfrac{x_r - x_1}{2x_m} \times 100$
51	51.5	52	48.6154	1.03
51.5	51.75	52	51.5575	0.48
51.5	51.625	51.75	50.4770	0.25
51.5	51.5625	51.625	49.6421	0.12
51.5625	51.5938	51.625	50.0845	0.06

(c) For $x = 3.5$, $x + x^x = 83.7$
For $x = 3.6$, $x + x^x = 104.2$

x_1	x_m	x_r	$f(x_m)$	$\dfrac{x_r - x_1}{2x_m} \times 100$
3.5	3.55	3.6	93.3571	1.41
3.55	3.575	3.6	98.6271	0.70
3.575	3.5875	3.6	101.3823	0.35
3.575	3.5813	3.5875	100.0055	0.17
3.575	3.5782	3.5813	99.3247	0.09

3. For each of the examples above, estimate how many iterations should have been needed to reach the specified accuracy.

(a) $E_i = 0.05$ in absolute terms, or
$E_i = 0.05 \times 100/5.65 = 0.88$ as a percentage of the estimate of the solution.

Also, $E_r = 0.1\%$. Therefore $N = 3.32 \log(E_i/E_r) = 3.14 = 4$ to the next higher integer. (Only a whole number of iterations can be performed.)

(b) $E_i = 0.5 \times 100/51.5 = 0.97\%$. Again, $E_r = 0.1\%$.
Therefore $N = 3.28 \Rightarrow 4$.

(c) $E_i = 0.05 \times 100/3.55 = 1.41\%$. Again, $E_r = 0.1\%$.

Therefore $N = 3.82 \Rightarrow 4$.
Thus, four iterations in each case should have been enough – and were!

4. Solve each of the equations in Example 1 by simple iteration. Check that the convergence criterion is satisfied for the particular arrangement of the equation used.

(a) Try $x = \{1000 - 2/x^4\}^{1/4} = F(x)$, for which

$$F'(x) = \frac{2}{x^5(1000 - 2/x^4)^{3/4}}$$

(Check this!). Therefore $F'(5.65) = 1.95 \times 10^{-6}$, i.e. $|F'(5.65)| \ll 1$ and convergence should be very rapid:

n	x_n
1	5.65
2	5.623 410 493
3	5.623 410 440
4	5.623 410 440

It is.

Note (1) The error in x_1 turns out to be $e_1 = 0.026\,589\,559\,8$, while $e_2 = 0.000\,000\,052\,6$. Thus, $e_2/e_1 = 1.98 \times 10^{-6} \approx F'(x_1)$, and (2.21) is confirmed.
(2) Calculation shows that $|F'(x)| < 1$ for $0.41 < x < \infty$, so that simple iteration will converge for any sensible first estimate.

(b) Try $x = 50/\sin x$, for which
$$F'(x) = (50 \cos x)/\sin^2 x$$
Therefore $F'(51.5) = 18.5$ and convergence is *not* guaranteed:

n	x_n
1	51.5
2	52.297
3	117.30
4	−57.37

and is not obtained.

Try $x = \sin^{-1}(50/x) = F(x)$, for which
$$F'(x) = \frac{-50/x^2}{\sqrt{(1 - 2500/x^2)}}$$

Therefore $F'(51.5) = -0.079$ and the convergence criterion is satisfied.

There is a minor problem: for $x = 51.5$ a calculator with trigonometrical functions or a book of mathematical tables gives a value of 1.3289 for $\sin^{-1}(50/x)$, i.e. for the next estimate of x. However, as was shown during the discussion of Example 1b, the solution has to be greater than 50.

Therefore add 16π to obtain 51.5943. This has the same sine.

n	x_n
1	51.5
2	51.5943
3	51.5870
4	51.5876
5	51.5875

Now that we know the answer ($S = 51.5875$), we can see that $e_1 = -0.0875$, $e_2 = 0.0068$ and $e_3 = -0.0005$. The errors are alternating in sign and reducing by a factor of 0.078, consistent with $F'(51.5) = -0.079$. Equation (2.21) triumphs again.

(c) Try $x = 100 - x^x = F(x)$, for which
$$F'(x) = -x^x(\ln x + 1)$$

With $x \approx 3\frac{1}{2}$, $|F'(x)|$ will obviously not be less than unity.
Try $x = (100 - x)^{1/x}$, for which

$$F'(x) = -(100-x)^{(1-x)/x}\left(\frac{(100-x)\ln(100-x)}{x^2} + \frac{1}{x}\right)$$

(Verify this!). Therefore $F'(3.5) = -1.38$. Better, but it still does not converge. (Do it: successive estimates are $3.5, 3.69, 3.45, 3.76, 3.37, \ldots$.)

Try
$$x^x = 100 - x$$
$$x \ln x = \ln(100-x)$$
$$x = \ln(100-x)/\ln x = F(x)$$

for which

$$F'(x) = -\frac{1}{(100-x)\ln x} - \frac{\ln(100-x)}{x(\ln x)^2}$$

Therefore $F'(3.5) = -0.84$ and convergence is assured: x converges slowly (the error reducing by a factor of about 0.84 at each step) to 3.58128 in 57 iterations.

What is the range of x for which $|F'(x)| < 1$? Trial calculations show that it is about $3.276 < x < 99.999$. [For $x \geq 100$, $\ln(100-x)$ blows up.] However, this does not mean that convergence is guaranteed for any starting estimate in this interval. The convergence criterion (see Section 2.5) is that $|F'(S)| < 1$, where S is the solution. The trouble is that we do not know the solution before we start, so we have to use the best estimate available, which is 3.5 in this case.

In fact, convergence will *not* occur for all x in the interval (3.276, 99.999). For example, if $x_1 = 48$, then $x_2 = 1.02$, which is outside the range of convergence; $x_3 = 224.5$; and x_4 is undefined, because $\ln(-124.5)$ is undefined.

5. Improve the rate of convergence of Example 4c with the use of Aitken's extrapolation.

n	x_n		
1	3.5		
2	3.64757	$\dfrac{x_3 x_1 - x_2^2}{x_3 - 2x_2 + x_1}$	$= 3.58213$
3	3.52998		
4	3.58213		
5	3.58060	$\dfrac{x_6 x_4 - x_5^2}{x_6 - 2x_5 + x_4}$	$= 3.58128$
6	3.58181		

and the solution has converged with about one-tenth of the previous computational effort.

6. Investigate whether damped iteration can cause the solution of Example 4c to converge even more quickly than in Example 5.

WORKED EXAMPLES

$$x_{n+1} = \lambda F(x_n) + (1 - \lambda)x_n = G(x_n)$$

where $F(x) = \ln(100 - x)/\ln x$. Now,

$$G'(x) = \lambda F'(x) + 1 - \lambda$$

and $x_{n+1} = G(x_n)$ will converge if $|G'(S)| < 1$, i.e. if λ is in the interval $(0, 2/(1 - F'(S))$.

Using $S \approx 3.5$, $F'(S) \approx -0.84$ (see Example 5); therefore $0 < \lambda < 1.087$. The best value of λ – the one which makes $x_{n+1} = G(x_n)$ converge the most quickly – is the value which makes $G'(S) = 0$. The best estimate of $G'(S)$ at any stage in the iterations is $G'(x_n)$; setting $G'(x_n)$ to zero leads to

$$\lambda = \frac{1}{1 - F'(x_n)}$$

The initial estimate, x_1, is 3.5, therefore $\lambda = -0.5435$. With this value of λ the iteration $x_{n+1} = G(x_n)$ yields 3.5, 3.580 21, 3.581 25, 3.581 27, 3.581 28, 3.581 28. (The extra iteration is necessary because we would not know that 3.581 28 is the solution until it has appeared twice.) So convergence is obtained in five iterations, i.e. five evaluations of $F(x)$. As it happens, the same number of function evaluations are required as in Example 5, but the subsidiary calculations are slightly less time-consuming with damped iteration.

7. Investigate the use of damped iteration on $x = 100 - x^x = F(x)$, which (see Example 4c) is rapidly divergent in its present form.

$F'(x) = -x^x(\ln x + 1)$, and $F'(3.5) = -180.7$. Therefore convergence is obtained with $0 < \lambda < 2/\{1 - F'(3.5)\} = 0.011\,01$ and the best value of λ is $1/\{1 - F'(3.5)\} = 0.005\,50$. With this value of λ,

$$x_{n+1} = \lambda(100 - x_n^{x_n}) + (1 - \lambda)x_n = G(x_n)$$

converges as follows: 3.5, 3.589 59, 3.579 41, 3.581 67, 3.581 19, 3.581 29, 3.581 27, 3.581 28.

Now we know that the true value of S is 3.581 28, we can find $F'(x) = -219.4$; so the actual range of values of λ for which $x = G(x)$ converges is $0 < \lambda < 0.009\,07$, with an optimum value of 0.004 54. Using this optimum, the successive estimates of S are 3.5, 3.573 95, 3.581 22, 3.581 28. Convergence is obtained in three iterations – but we needed to know the solution first! The rate of convergence using λ based on $S \approx 3.5$ is obviously acceptable.

8. Solve

$$P(x) = x^4 - 10x^3 + 35x^2 - 50x + 20 = 0$$

by Newton's method and synthetic division.

There are four roots. Assuming they are real, they must all be positive, since there are no sign repetitions and therefore no negative roots. A first estimate of the largest root is $-(-10)/1 = 10$. (If the largest root is by far the largest, then this will be a good first estimate. If not, it is still better than just a guess.) So:

$$
\begin{array}{r|rrrrr}
x_1 = 10 & 1 & -10 & 35 & -50 & 20 \\
 & & 10 & 0 & 350 & 3000 \\
\hline
10 & 1 & 0 & 35 & 300 & 3020 = P(10) \\
 & & 10 & 100 & 1350 & \\
\hline
 & 1 & 10 & 135 & 1650 = P'(10) &
\end{array}
$$

Therefore $x_2 = 10 - 3020/1650 = 8.17$. Repeating the process, and adopting an even more compact style, we obtain

$$
\begin{array}{r|rrrrr}
x_2 = 8.17 & 1 & -10 & 35 & -50 & 20 \\
8.17 & 1 & -1.83 & 20.05 & 113.78 & 949.53 = P(8.17) \\
\hline
 & 1 & 6.34 & 71.84 & 700.67 = P'(8.17) &
\end{array}
$$

Therefore $x_3 = 6.81$. Successive estimates (a programmable calculator makes life a lot easier) are 5.83, 5.13, 4.67, 4.44, 4.37 and 4.37. (These figures are rounded to two decimal places, but were calculated to 12 significant figures). Using $x = 4.37$:

$$
\begin{array}{r|rrrrr}
4.37 & 1 & -10 & 35 & -50 & 20 \\
\hline
 & 1 & -5.63 & 10.40 & -4.57 & 0.05 = P(4.37)
\end{array}
$$

This is the best answer to two decimal places. A better solution is 4.3671, for which $P(x) = -0.0000249$ and the reduced polynomial (the result of dividing $P(x)$ by $(x - 4.3671)$ is

$$Q(x) = x^3 - 5.6329x^2 + 10.4006x - 4.5797$$

Now we see that $x_1 = 10$ was not a very good first estimate! The only other real root turns out to be 0.6329.

Problems

1. Use common sense, simplification, graphing or intuition to obtain a first estimate of a solution of the following equations:

(a) $x^3 + 1/x^3 = 10$
(b) $x + \log x = 50$
(c) $x + e^x = 50$

(d) $x^3 - \cos x - 20 = 0$
(e) $x^{3.9} - 3x^{1.9} + 2 = 0$
(f) $x^3 - 3x^2 + 3x - 1.5 = 0$
(g) $\cos x \cosh x = 1$
(h) $x + \cosh x = 50$
(i) $x + \cos x = 0.1$

2. For each of the equations in Question 1, find an interval Δx within which the solution of the equation is located. Then solve the equation by interval halving, performing sufficient iterations to obtain an error of 0.1% or less.

3. Rearrange each of the equations in Question 1 into the form $x = F(x)$, suitable for solution by simple iteration. For the estimate of the solution obtained in Question 1, determine whether the convergence criterion for simple iteration is satisfied. If so, find the range of values of x for which the criterion is satisfied, and solve the equation by simple iteration.

4. Solve the equation

$$\sin x + \sin^{-1} x = 1$$

by rearranging it as

$$x = \sin(1 - \sin x)$$

and using simple iteration. How many iterations are required to obtain the solution to an error of 0.01% or less?

Next, use Aitken's extrapolation after each set of three iterations. How many iterations are now required to obtain the same accuracy?

5. Consider the equation

$$x = x^{1/2} - 0.2499$$

of which a solution is $x = 0.2601$. Estimate the number of iterations, starting with $x_1 = 0.3$, which are required to obtain the solution to within 0.01%.

Solve the equation, using Aitken's extrapolation after each set of three iterations. How many iterations are now required to obtain the desired accuracy?

6. Consider the function

$$f(x) = e^{-x} \sin x + 0.2x - 0.521$$

(a) Sketch the function, and hence locate approximately the value or values of x at which $f(x) = 0$.

(b) Solve the equation $f(x) = 0$ by interval halving, to an accuracy of 0.01%.

(c) Solve the equation $f(x) = 0$ by simple iteration, to the same accuracy.

7. Consider the equation

$$f(x) = x^3 + 2x^2 + 10x - 20 = 0$$

(a) Verify that there is only one positive root of this equation; that it lies between 1 and 2; and that $x = 1.3$ is an estimate of it.

(b) Solve the equation by the interval-halving method, performing sufficient iterations to obtain the solution to within 0.01%.

(c) Show that the equation, rewritten as

$$x = F(x) = \frac{20}{x^2 + 2x + 10}$$

can be solved by simple iteration (i.e. that the convergence criterion is satisfied), and find the solution. How many iterations are required to obtain an accuracy of 0.01%?

(d) Find $F'(1.3)$ and hence estimate the optimum damping factor λ for use in the damped iteration process

$$x = G(x) = \lambda F(x) + (1 - \lambda)x$$

Find the solution with this value of λ. How many iterations are now required to obtain 0.01% accuracy?

(e) Consider, in turn, each of the following versions of the equation:

(i) $x = \dfrac{20 - 2x^2 - x^3}{10}$

(ii) $x = \dfrac{20 - 10x - 2x^2}{x^2}$

(iii) $x = \dfrac{20 - 10x - x^3}{2x}$

(iv) $x = (20 - 10x - 2x^2)^{1/3}$

By calculating $F'(1.3)$, show that these equations do *not* satisfy the convergence criterion. Find a range of values of the damping factor for which the damped iteration process *will* converge, and estimate the optimum value of λ.

(f) Starting with $x = 1.3$, solve the original equation by Newton's method, using synthetic division. Verify that two iterations are sufficient to obtain the solution to within 0.01%.

8. For equation (h) of Question 1, an obvious rearrangement is

$$x = 50 - \cosh x = F(x)$$

and a first estimate of the solution is about 4.

Show that $F'(4) \approx -27$ and therefore that simple iteration is not guaranteed to converge. Confirm that simple iteration does not, in fact, converge.

Show that, using damped iteration, the process defined by

$$x_{n+1} = \lambda F(x_n) + (1 - \lambda)x_n$$

will converge for equation (h) if $\lambda \approx 0.035$, the value corresponding to $F'(4) = -27$. Verify that the solution can thus be obtained within 0.1%, starting from $x_1 = 4$, in nine iterations.

Given that the solution is 4.5105 (to five significant figures), show that $\lambda_{opt} = 0.0215$ and that, using this value of λ, damped iteration converges from $x_1 = 4$ to an accuracy of better than 0.1% in three iterations.

9. For equation (i) of Question 1, an obvious rearrangement is

$$x = 0.1 - \cos x = F(x)$$

and a first estimate of the solution is about -0.7.

Show that $F'(-0.7) \approx -0.644$, and confirm that simple iteration converges from $x_1 = -0.7$ to within 0.1% in eight iterations. Having obtained the solution, compute the error at each iteration, and hence the ratio of the error at each iteration to the error at the previous iteration.

Given that the solution is -0.6785, show that $\lambda_{opt} = 0.6143$, and that using this value of λ, damped iteration converges from $x_1 = -0.7$ to better than 0.1% in one iteration.

10. The value of λ for use in the method of damped iteration can be estimated, without knowing the true solution, by using $F'(x_n)$ as an approximation to $F'(S)$; λ can then be recalculated after each iteration from

$$\lambda_{n+1} = \frac{1}{1 - F'(x_n)}$$

Apply this idea to equation (i) of Question 1, for which $F'(x) = \sin x$. Starting with $\lambda = 1$ and $x = -0.7$, show that damped iteration with successive adjustment of λ converges to an accuracy of better than 0.01% in two iterations.

11. Use the method of Question 9 to solve the equation in Question 5. Starting with $x_1 = 3$, how many iterations are required to obtain the answer to within 0.01%? What is the final value of λ?

12. Solve the equation

$$x + e^x = A$$

with (a) $A = 50$, (b) $A = 5$ and (c) $A = 1.1$. For each value of A, apply simple iteration to the equations in the two arrangements

(i) $x = A - e^x$

(ii) $x = \ln(A - x)$

using suitable initial estimates of S in each case. Determine whether the iterations converge, and, if so, find the solution. What is the number of iterations needed to obtain the solution to an accuracy of 0.1%?

Now use damped iteration with automatic adjustment of λ, as described in Question 9, to solve each arrangement of the equation for each value of A. Using the same initial estimates as before, determine the number of iterations now needed to obtain 0.01% accuracy.

13. Solve the equation

$$e^{at} - at - b = 0$$

for $a = 0.4$, $b = 9$, using (i) interval halving, (ii) simple iteration and (iii) Newton's method. Note that there may be more than one root.

14. The van der Waals equation

$$(p + a/v^2)(v - b) = R_0 T$$

is, under certain conditions, a reasonably accurate relationship between the pressure p, volume v and absolute temperature T of one mole of a gas. The quantities a and b are constants whose values are different for different gases, and R_0 is the universal gas constant. $R_0 = RM$, where R is the gas constant for a particular gas of molecular weight M.

Write a program for a computer or a programmable calculator which will accept values of a, b and M and then, for given values for any two of p, v and T, will compute the value of the third gas property.

Use the 'ideal gas equation'

$$pv = R_0 T$$

to furnish an initial estimate, and take $R_0 = 8.3143$ kJ kmol^{-1} K^{-1}.

The following are values of a, b and M for some gases. Use them to test your program, and compare your results with values that may be found in tables of gas properties. The extent to which your results differ from the tabulated values will be a measure of the reliability of the van der Waals equation (not to mention your program).

Gas	a (kPa m^6 kmol^{-2})	b (m^3 kmol^{-1})	M (kg kmol^{-1})
air	135.8	0.0364	28.97
carbon dioxide	364.3	0.0427	44.01
helium	3.41	0.0234	4.003
propane	931.5	0.0900	44.09

PROBLEMS

15. The buckling load on a centrally loaded column is given by
$$P_{cr} = k^2 EI$$
where E is Young's modulus and I is the minimum moment of inertia of the cross section.

For a column that is built-in at one end and pinned at the other, k is obtained from
$$\frac{\sin kL}{k} - L \cos kL = 0$$
where L is the length of the column. Show that, for such a column,
$$P_{cr} = 20.19 EI/L^2$$

16. Given that the stress tensor at a point is
$$\begin{pmatrix} \sigma_x & \tau_{xy} & \tau_{xz} \\ \tau_{yx} & \sigma_y & \tau_{yz} \\ \tau_{zx} & \tau_{zy} & \sigma_z \end{pmatrix}$$
the principal stresses at that point are the roots of
$$\sigma^3 - I_1 \sigma^2 + I_2 \sigma - I_3 = 0$$
in which
$$I_1 = \sigma_x + \sigma_y + \sigma_z$$
$$I_2 = \begin{vmatrix} \sigma_x & \tau_{xy} \\ \tau_{yx} & \sigma_y \end{vmatrix} + \begin{vmatrix} \sigma_y & \tau_{yz} \\ \tau_{zy} & \sigma_z \end{vmatrix} + \begin{vmatrix} \sigma_z & \tau_{zx} \\ \tau_{xz} & \sigma_x \end{vmatrix}$$
and
$$I_3 = \begin{vmatrix} \sigma_x & \tau_{xy} & \tau_{xz} \\ \tau_{yx} & \sigma_y & \tau_{yz} \\ \tau_{zx} & \tau_{zy} & \sigma_z \end{vmatrix}$$
Determine the principal stresses at a point where the stress tensor is
$$\begin{vmatrix} 11.47 & 3.81 & 0.55 \\ 3.81 & 7.12 & 5.14 \\ 0.55 & 5.14 & 4.68 \end{vmatrix} \text{ MPa}$$

17. Find the real roots of
$$x^4 + 5.6x^3 - 21.61x^2 - 12.896x + 51.422 = 0$$

18. Find the real roots of
$$x^4 - 3.2x^3 + 1.21x^2 + 0.21x + 4.41 = 0$$

19. Find the real roots of

$$x^4 - 6x^3 + 13x^2 - 12x + 3.99 = 0$$

20. Find the complex roots, if any, of the equations in Problems 17–19.

3

Simultaneous equations

3.1 Introduction

In Chapter 2 methods for solving an algebraic or transcendental equation containing a single unknown were discussed. We now turn to *systems of equations*, and in particular to systems of n equations involving n unknowns*.

Problems involving systems of equations arise in all branches of science and engineering: for example, electric circuit analysis, radiation heat transfer and stress analysis in structures. Systems of equations are also relevant to some methods for the solution of differential equations.

The equations may be *linear* or *non-linear*. The methods for solving linear systems are far more highly developed than those for non-linear systems. For this reason more attention will be paid to the former. However, because of their importance, non-linear equations will also be considered, if only briefly.

In general, there will be exactly one set of values of the n unknowns which will satisfy the n equations. This is not necessarily the case, however. For example, it is apparent that the system

$$\begin{aligned} 2x_1 + 2x_2 &= 2 \\ 2x_1 + 2x_2 &= 3 \end{aligned} \qquad (3.1)$$

cannot be satisfied for *any* choice of x_1 and x_2; while the system

$$\begin{aligned} x_1 + x_2 &= 2 \\ 2x_1 + 2x_2 &= 4 \end{aligned} \qquad (3.2)$$

can be satisfied for an infinite set of values (but not for all possible sets!).

---

* If there are more independent unknowns than equations, then there is insufficient information to enable a solution to be found. On the other hand, if there are more linearly independent equations than unknowns, then not all of the equations can be satisfied exactly. However, values of the unknowns can be found which allow the equations to be satisfied as well as possible. Least squares analysis is one technique for doing this.

Systems like (3.1) and (3.2) are said to be *singular*. It can be seen from these examples that systems of linear equations do not possess a unique solution when the determinant formed from the coefficients of the unknowns is zero. A situation such as that occurring with (3.2) is due to the fact that the equations are not *linearly independent*: one of the equations can be formed by a suitable linear combination of some (in this case, one) of the other equations. That equation does not provide new information, and the system is therefore underdetermined – there are not n equations for the n unknowns.

Linear systems can be solved either by *direct* methods or by *iterative* methods. *Direct methods* enable the exact solution to be found in a finite and predictable number of operations. The number depends on n, the *order* of the system. For the commonest direct method, that of elimination, the number is roughly proportional to n^3. *Iterative methods*, which are essentially extensions of some of the iterative techniques described in the previous chapter, will again allow only an approximate solution to be found (although the approximation can, as before, be made as good as we wish within the available limits of accuracy). The number of iterations is almost independent of the order of the system, depending far more strongly on the quality of the initial estimate of the solution and on the values of the coefficients in the equations (which determine the rate of convergence per iteration). The number of operations *per iteration* is proportional to n^2. Thus, for small systems direct methods are faster, while for large systems – and it is not uncommon for a system of 1000 or more equations to be encountered (for example, in the numerical solution of elliptic partial differential equations, discussed in Ch. 6) – iterative methods are preferable (if they are convergent – see later in this chapter). It is not possible to place a clear line of demarcation between the two methods, but it *usually* appears better, if possible, to use iterative methods for a general system if n is greater than about 100.

For some special systems, e.g. tridiagonal systems and pentadiagonal systems, in which the only non-zero elements are located on the three or five leading diagonals, respectively, special direct methods are normally the best for systems of any size. These methods are described below.

These comments about the relative merits of direct and iterative methods must be regarded as generalizations at best: particular procedures are appropriate under particular circumstances, which students will encounter in more-advanced courses in numerical methods than those for which this book is suitable. Other factors, such as the available memory on the computer, or whether the computer is a vector processor or a parallel processor, are also relevant. It is merely intended here to suggest that direct methods for large systems are generally time-consuming and that iterative methods may then be preferable.

3.2 Elimination methods

In a general linear system, each of the n equations may contain all of the n unknowns in a linear combination. Elimination methods are based on forming linear combinations of the equations in such a way that some of the unknowns are eliminated from some of the equations until there is one equation with only one unknown, which can then be found. The remaining unknowns are determined by a continuation of the process.

Consider, for example, the system

$$2x_1 + x_2 + x_3 = 9 \qquad (3.3a)$$
$$3x_1 + x_2 - x_3 = 10 \qquad (3.3b)$$
$$2x_1 - 2x_2 + 2x_3 = 4 \qquad (3.3c)$$

If (b) is replaced by the result of subtracting two times (b) from three times (a), and if (c) is replaced by the result of subtracting (c) from (a), we obtain

$$2x_1 + x_2 + x_3 = 9 \qquad (3.3d)$$

$3(a) - 2(b)$:
$$x_2 + 5x_3 = 7 \qquad (3.3e)$$

$(a) - (c)$:
$$3x_2 - x_3 = 5 \qquad (3.3f)$$

If now (f) is replaced by the result of subtracting three times (e) from (f) we obtain

$$2x_1 + x_2 + x_3 = 9 \qquad (3.3g)$$
$$x_2 + 5x_3 = 7 \qquad (3.3h)$$

$(f) - 3(e)$:
$$-16x_3 = -16 \qquad (3.3i)$$

Equation (3.3i) leads immediately to $x_3 = 1$. Then (h) shows that $x_2 = 2$ and (g) yields $x_1 = 3$. The unknown x_1 was eliminated from (e), and both x_1 and x_2 were eliminated from (i). After finding x_3, its value was substituted back into (h), and x_2 and x_3 were substituted into (g). The processes of *elimination* and *back substitution* are the essential features of these direct methods.

The foregoing equations can be written as matrix equations. For example, (3.3a–c) are equivalent to

$$\begin{pmatrix} 2 & 1 & 1 \\ 3 & 1 & -1 \\ 2 & -2 & 2 \end{pmatrix} \begin{pmatrix} x_1 \\ x_2 \\ x_3 \end{pmatrix} = \begin{pmatrix} 9 \\ 10 \\ 4 \end{pmatrix} \qquad (3.4)$$

while (3.3g–i) are equivalent to

$$\begin{pmatrix} 2 & 1 & 1 \\ 0 & 1 & 5 \\ 0 & 0 & -16 \end{pmatrix} \begin{pmatrix} x_1 \\ x_2 \\ x_3 \end{pmatrix} \begin{pmatrix} 9 \\ 7 \\ -16 \end{pmatrix}$$

The operations performed on the equations are clearly equivalent to operations which could have been performed on the matrix of the coefficients and on the vector of the right-hand sides.

Finally, we observe that we need not actually write down the vector of unknowns: we could operate instead with what is called the *augmented* matrix

$$\begin{pmatrix} 2 & 1 & 1 & 9 \\ 3 & 1 & -1 & 10 \\ 2 & -2 & 2 & 4 \end{pmatrix} \qquad (3.5)$$

formed from the matrix and vector of (3.4). Starting with this matrix, we seek operations which will yield

$$\begin{pmatrix} 2 & 1 & 1 & 9 \\ 0 & 1 & 5 & 7 \\ 0 & 0 & -16 & -16 \end{pmatrix}$$

We would like a method which can, when required, be implemented on a computer. When these calculations are performed by hand, advantage can be taken of particular values of the coefficients to simplify the working. For example, it was easy for a human to see that (3.3e) and (3.3f) should be combined by multiplying the former by 3 and adding. However, when the operations are to be automated, we need a systematic process which, although perhaps involving some extra effort which would not be performed by people, does not require a knowledge of any special relationships between the coefficients. The objective should still be to find a *triangular* system like (3.3g–i) which can be readily solved and which has the same solution as the original equations.

This can be achieved in the following way, using (3.5) as an example:

Original matrix:
$$\begin{pmatrix} 2 & 1 & 1 & 9 \\ 3 & 1 & -1 & 10 \\ 2 & -2 & 2 & 4 \end{pmatrix}$$

Divide 1st row by 2 (the 1st diagonal value) to put a '1' on the diagonal; subtract 3 times the *new* first row from the second row, and 2 times the *new* first row from the third row, to put '0's in the subdiagonal positions of the first column:
$$\begin{pmatrix} 1 & \tfrac{1}{2} & \tfrac{1}{2} & 4\tfrac{1}{2} \\ 0 & -\tfrac{1}{2} & -2\tfrac{1}{2} & -3\tfrac{1}{2} \\ 0 & -3 & 1 & -5 \end{pmatrix}$$

Divide 2nd row by $-\tfrac{1}{2}$ (the current 2nd diagonal value) to put a '1' on the diagonal; subtract (-3) times *the new* second row from the third row, to put a '0' in the subdiagonal position of the second column:
$$\begin{pmatrix} 1 & \tfrac{1}{2} & \tfrac{1}{2} & 4\tfrac{1}{2} \\ 0 & 1 & 5 & 7 \\ 0 & 0 & 16 & 16 \end{pmatrix}$$

Divide 3rd row by 16 (the current 3rd diagonal value) to put a '1' on the diagonal:
$$\begin{pmatrix} 1 & \frac{1}{2} & \frac{1}{2} & 4\frac{1}{2} \\ 0 & 1 & 5 & 7 \\ 0 & 0 & 1 & 1 \end{pmatrix}$$

The system of equations represented by this matrix is triangular, and can easily be solved. Its derivation was purely mechanical, no advantage being taken of special values of the coefficients to achieve short cuts. We now consider the generalization of this process.

3.3 Gaussian elimination

It is most unlikely that students will ever need to write their own elimination routines: these routines are readily available in computer centre libraries and it is pointless to duplicate existing procedures. Nevertheless, they should understand the programs they are using, to ensure they are using them to the best advantage and in the appropriate circumstances. The most commonly used elimination routine is that associated with the name of Gauss.

Consider the system of n equations

$$\sum_{j=1}^{n} a_{ij}x_j = a_{i,n+1}, \quad i = 1, 2, \ldots, n \tag{3.6}$$

for which the augmented matrix is the $n \times (n + 1)$ matrix

$$\begin{bmatrix} a_{11} & a_{12} & a_{13} & \cdots & & a_{1n} & a_{1,n+1} \\ a_{21} & a_{22} & a_{23} & \cdots & & & a_{2,n+1} \\ \vdots & \vdots & \vdots & & & & \vdots \\ a_{i1} & a_{i2} & a_{i3} & \cdots & a_{ii} & \cdots & a_{i,n+1} \\ \vdots & \vdots & \vdots & & & & \vdots \\ a_{n1} & a_{n2} & a_{n3} & \cdots & & a_{nn} & a_{n,n+1} \end{bmatrix} \tag{3.7}$$

Operations are performed which have the following effects:

(a) Replacement of each diagonal element a_{ii} by unity;
(b) Replacement of each element *below* the diagonal by zero;
(c) Replacement of all other elements by the values resulting from the operations to achieve (a) and (b).

The final result will be the replacement of (3.7) by

$$\begin{bmatrix} 1 & a'_{12} & a'_{13} & \cdots & & a'_{1n} & a'_{1,n+1} \\ 0 & 1 & a'_{23} & \cdots & & & a'_{2,n+1} \\ \vdots & \vdots & \vdots & & & & \vdots \\ 0 & 0 & 0 & \cdots & 1 & \cdots & a'_{i,n+1} \\ \vdots & \vdots & \vdots & & & & \vdots \\ 0 & 0 & 0 & \cdots & & 1 & a'_{n,n+1} \end{bmatrix} \tag{3.8}$$

We see that we only need to compute slightly less than half the elements in (3.8): we know that all the diagonal elements are unity, and that all the subdiagonal elements are zero. Advantage is taken of this in the following algorithm:

$$\begin{aligned}&\text{For } k = 1, 2, \ldots, n\text{:} \\ &\quad \text{For } j = k + 1, k + 2, \ldots, n + 1\text{:} \\ &\quad\quad a_{kj} \leftarrow a_{kj}/a_{kk}; \\ &\quad \text{For } i = k + 1, k + 2, \ldots, n \text{ (but not when } k = n\text{):} \\ &\quad\quad \text{For } j = k + 1, k + 2, \ldots, n + 1\text{:} \\ &\quad\quad\quad a_{ij} \leftarrow a_{ij} - a_{ik}a_{kj}.\end{aligned} \quad (3.9)$$

where the left-pointing arrow ($\leftarrow$) is to be read 'is replaced by'.

Expressed in words, this algorithm performs the following steps.

(a) Select each of the rows in turn; number the selected row 'k'.

(b) Divide each element a_{kj} to the right of the diagonal element a_{kk} by a_{kk}. We need not include elements to the left of the diagonal because they would become zero in any case if we were to perform the subsequent operations on them; and we need not include a_{kk} itself: we simply assume hereafter that it is unity. That is, we do not do the arithmetic for those elements for which we know the result – zero or unity, as the case may be. For this reason, these elements retain their original values in the examples below, because we have not, in fact, found new values for them; but in effect, all leading diagonal elements are eventually unity, and all elements below the leading diagonal are eventually zero.

(c) For each row k, select in turn each of the other rows from the range $k + 1$ to n; number the selected row 'i'.

(d) In each row i, select in turn each of the elements from the columns in the range from $k + 1$ to $n + 1$; number the selected column 'j', i.e. the selected element is a_{ij}.

(e) Replace a_{ij} by $a_{ij} - a_{ik}a_{kj}$. This would set a_{ij} to zero for $j = k$; and for $j < k$, a_{ij} would have become zero when the smaller values of k were used. We have therefore omitted these values of j to avoid doing unnecessary work.

Let us apply the algorithm to the matrix (3.5):

$$\begin{pmatrix} 2 & 1 & 1 & 9 \\ 3 & 1 & -1 & 10 \\ 2 & -2 & 2 & 4 \end{pmatrix}$$

(1) Set $k = 1$. Divide each element in the first row after the first element by $a_{11} = 2$. This produces

$$\begin{pmatrix} 2 & \tfrac{1}{2} & \tfrac{1}{2} & 4\tfrac{1}{2} \\ 3 & 1 & -1 & 10 \\ 2 & -2 & 2 & 4 \end{pmatrix}$$

(2) Set $i = 2$.
(3) Set $j = 2, 3$ and 4 in turn, and compute

$$a_{22} = a_{22} - a_{21}a_{12} = 1 - (3)(\tfrac{1}{2}) = -\tfrac{1}{2}$$
$$a_{23} = a_{23} - a_{21}a_{13} = -1 - (3)(\tfrac{1}{2}) = -2\tfrac{1}{2}$$
$$a_{24} = a_{24} - a_{21}a_{14} = 10 - (3)(4\tfrac{1}{2}) = -3\tfrac{1}{2}$$

(4) Set $i = 3$.
(5) Set $j = 2, 3$ and 4 in turn, and compute

$$a_{32} = a_{32} - a_{31}a_{12} = -2 - (2)(\tfrac{1}{2}) = -3$$
$$a_{33} = a_{33} - a_{31}a_{13} = 2 - (2)(\tfrac{1}{2}) = 1$$
$$a_{34} = a_{34} - a_{31}a_{14} = 4 - (2)(4\tfrac{1}{2}) = -5$$

The matrix is now

$$\begin{pmatrix} 2 & \tfrac{1}{2} & \tfrac{1}{2} & 4\tfrac{1}{2} \\ 3 & -\tfrac{1}{2} & -2\tfrac{1}{2} & -3\tfrac{1}{2} \\ 2 & -3 & 1 & -5 \end{pmatrix}$$

(6) Set $k = 2$. Divide each element in the second row after the second element by $a_{22} = -\tfrac{1}{2}$. This produces

$$\begin{pmatrix} 2 & \tfrac{1}{2} & \tfrac{1}{2} & 4\tfrac{1}{2} \\ 3 & -\tfrac{1}{2} & 5 & 7 \\ 2 & -3 & 1 & -5 \end{pmatrix}$$

(7) Set $i = 3$.
(8) Set $j = 3$ and 4 in turn, and compute

$$a_{33} = a_{33} - a_{32}a_{23} = 1 - (-3)(5) = 16$$
$$a_{34} = a_{34} - a_{32}a_{24} = -5 - (-3)(7) = 16$$

The matrix is now

$$\begin{pmatrix} 2 & \tfrac{1}{2} & \tfrac{1}{2} & 4\tfrac{1}{2} \\ 3 & -\tfrac{1}{2} & 5 & 7 \\ 2 & -3 & 16 & 16 \end{pmatrix}$$

(9) Set $k = 3$. Divide each element in the third row after the third element by $a_{33} = 16$. This produces

$$\begin{pmatrix} 2 & \tfrac{1}{2} & \tfrac{1}{2} & 4\tfrac{1}{2} \\ 3 & -\tfrac{1}{2} & 5 & 7 \\ 2 & -3 & 16 & 1 \end{pmatrix}$$

(10) With $k = 3$, there are no possible values of i. The algorithm [as described by (3.9)] is thus complete. Notice that the elements on and beneath the leading diagonal still have their original values – because we

have not changed them. But for clarity we can now insert into the matrix the diagonal elements (1) and the subdiagonal elements (0) which would have been computed – except that we saved ourselves the effort. The matrix is finally

$$\begin{pmatrix} 1 & \frac{1}{2} & \frac{1}{2} & 4\frac{1}{2} \\ 0 & 1 & 5 & 7 \\ 0 & 0 & 1 & 1 \end{pmatrix} \quad (3.10)$$

representing the system

$$\begin{aligned} x_1 + \tfrac{1}{2}x_2 + \tfrac{1}{2}x_3 &= 4\tfrac{1}{2} \\ x_2 + 5x_3 &= 7 \\ x_3 &= 1 \end{aligned} \quad (3.11)$$

The algorithm (3.9) thus produces a triangular system of equations, which can be readily solved in reverse order. The algorithm for this process is

$$\begin{aligned} &x_n = a_{n,n+1}; \\ &\text{For } i = n-1, n-2, \ldots, 1: \\ &\quad x_i \leftarrow a_{i,n+1}; \\ &\quad \text{For } j = i+1, i+2, \ldots, n: \\ &\quad\quad x_i \leftarrow x_i - a_{ij}x_j. \end{aligned} \quad (3.12)$$

The second part of the solution process finds the values of the unknowns in reverse order: x_n is given immediately by the last row of the (modified) augmented matrix, and the remaining values can then be found successively, since each earlier row of the matrix (representing each earlier equation) involves only one new unknown.

Applying this to the matrix (3.10), i.e. to the system of equations (3.11), we find that

$$\begin{aligned} x_3 &= 1 \\ x_2 &= 7 - (5)(1) = 2 \\ x_3 &= 4\tfrac{1}{2} - (\tfrac{1}{2})(2) - (\tfrac{1}{2})(1) = 3 \end{aligned}$$

The complete Gaussian elimination algorithm consists of (3.9) followed by (3.12). It can be implemented readily on a computer, as shown in Figure 3.1, which also includes a segment for checking the solution by multiplying the matrix of coefficients by the solution vector, to show that the original right-hand sides will be recovered. Note that in order to do so, the coefficient matrix A must be saved in a separate array (denoted here by B) because A is altered during the solution process.

More importantly, note that the 1s and 0s of step (10) above are not actually inserted into A: it is sufficient to continue *as though this has been done*, as is implied by the back-substitution steps in (3.12).

```
c
          dimension a(10,11), b(10,11), x(10), r(10)
c
          write (*,1)
   1      format(/' how many equations to be solved?')
          read (*,*) n
          np1 = n + 1
          write (*,2)
   2      format(' enter coefficients of augmented matrix, row by row')
          do 5 i = 1, n
             write (*,4) i
   4         format(' row', i3, ':')
   5         read (*,*) ( a(i,j), j=1, np1)
c
c         save a in b for checking the solution
c         (this section is not necessary for the solution, and
c         increases the storage requirements. It may be omitted)
c
          do 7 i = 1, n
          do 7 j = 1, np1
   7         b(i,j) = a(i,j)
c
c         elimination stage
c
          do 20 k = 1, n
             kp1 = k + 1
             do 10 j = kp1, np1
  10            a(k,j) = a(k,j) / a(k,k)
             if (k .eq. n) go to 20
             do 15 i = kp1, n
                do 15 j = kp1, np1
  15               a(i,j) = a(i,j) - a(i,k) * a(k,j)
  20      continue
c
c         back substitution stage
c
          x(n) = a(n,np1)
          do 30  k = 2, n
             i = np1 - k
             x(i) = a(i,np1)
             ip1= i + 1
             do 30 j = ip1, n
  30            x(i) = x(i) - a(i,j)*x(j)
c
c         write the solution
c
          write (*,40) (x(i), i = 1, n)
  40      format (' the solution is:' // ( 10f10.4 ))
c
c         check the solution
c
          do 50 i = 1, n
             r(i) = 0.
             do 50 j = 1, n
  50            r(i) = r(i) + b(i,j)*x(j)
c
          write (*,60) (r(i), i = 1, n)
  60      format(/' check on solution. computed r.h.s. vector is:'//
         1       (10f10.4))
          stop
          end
```

Figure 3.1 The solution of a linear system by Gaussian elimination.

A typical output from this program is:

```
how many equations to be solved?
3
enter coefficients of augmented matrix, row by row
row 1:
2   1   1   9
row 2:
3   1  -1  10
row 3:
2  -2   2   4
the solution is:
    3.0000     2.0000     1.0000
check on solution.  computed r.h.s.vector is:
    9.0000    10.0000     4.0000
```

3.4 Extensions to the basic algorithm

Situations can arise in which the basic algorithm (3.9), (3.12) will yield inaccurate values or even fail entirely.

Consider first the system

$$0.000\,100 x_1 + 1.00 x_2 = 1.00$$
$$1.00\ \ x_1 + 1.00 x_2 = 2.00 \qquad (3.13)$$

and suppose a computer or other calculating device is being used which can work to an accuracy of three significant figures. (Of course, all contemporary calculating devices can retain more figures than three; but the use of more figures only postpones the problem to be discussed, and the smaller number of figures makes the explanation simpler.) Step-by-step application of (3.9) yields

$$\begin{aligned}
a_{12} &\leftarrow a_{12}/a_{11} & &= (1.00)/(0.000\,100) = 10\,000 \\
a_{13} &\leftarrow a_{13}/a_{11} & &= (1.00)/(0.000\,100) = 10\,000 \\
a_{22} &\leftarrow a_{22} - a_{21}a_{12} & &= 1.00 - (1.00)(10\,000) \\
& & &= -9999 \\
& & &= -10\,000 \text{ (to three significant figures)} \\
a_{23} &\leftarrow a_{23} - a_{21}a_{13} & &= 2.00 - (1.00)(10\,000) \\
& & &= -9998 \\
& & &= -10\,000 \text{ (to three significant figures)} \\
a_{23} &\leftarrow a_{23}/a_{22} & &= (-10\,000)/(-10\,000) \\
& & &= 1.00
\end{aligned}$$

whence (3.12) yields

$$x_2 = a_{23} = 1.00$$
$$x_1 = a_{13} - a_{12}x_2 = 10\,000 - (10\,000)(1.00)$$
$$= 0.00$$

However, a more exact solution (retaining more significant figures) shows that the true solution is

$$x_1 = 1.00001\ldots$$
$$x_2 = 0.99999\ldots$$

The rounding process has therefore introduced a catastrophic error in x_1. On the other hand, if the same equations are written in the reverse order,

$$\begin{aligned} 1.00x_1 + 1.00x_2 &= 2.00 \\ 0.000100x_1 + 1.00x_2 &= 1.00 \end{aligned} \qquad (3.14)$$

it will be found that the elimination algorithm yields

$$x_1 = 1.00$$
$$x_2 = 1.00$$

which is a correct result to three significant figures.

The problem arises during the *normalization* of the first equation – its division by a_{11}. When this is a very small number, as in (3.13), large numbers are produced. To three significant figures,

$$10\,000 - 1 = 10\,000$$

and error is thereby introduced. It is eliminated by reordering the equations to avoid small numbers on the diagonal.

The situation becomes even more serious if a diagonal element about to be used to normalize an equation is exactly zero. The process immediately fails.

These difficulties are overcome in complete Gaussian elimination routines – including those to be found in computer centre libraries – by one of several methods, including rearranging the equations during computation. The book-keeping necessary to keep track of all of the rearrangements makes the program become more complex, and details are omitted here. It is only necessary to point out that, provided the system is not singular, a complete elimination routine will avoid these difficulties as far as possible, i.e. to the limits of accuracy of the computer. However, with a singular system, a stage must eventually be reached in which no rearrangement can be found which will prevent the use of a zero as a divisor. In this case, as we have seen before, the solution is either non-existent or not unique.

3.5 Operation count for the basic algorithm

We can estimate the number of operations involved in the basic algorithm (3.9), (3.12). We will count only multiplications and divisions, since these are more time-consuming on a computer or calculator than additions and subtractions are.

The normalization step involves n divisions for the first row, $(n-1)$ for the second, $(n-2)$ for the third, etc. There are thus

$$n + (n-1) + (n-2) + \cdots + 2 + 1 = n(n+1)/2$$

operations in this step.

The second stage of (3.9) involves n values of k; for each of these, there are $(n-k)$ values of i; for each of these in turn there are $(n-k+1)$ values of j; and, finally, for each of these, there is one multiplication operation. For each value of k there are thus $(n-k)(n-k+1) = (n^2 - 2nk + k^2 + n - k)$ operations, and hence the total for this stage is

$$\sum_{k=1}^{n} (n^2 - 2nk + k^2 + n - k)$$

operations. Now

$$\sum_{k=1}^{n} n^2 = n^3$$

$$\sum_{k=1}^{n} 2nk = 2n\{n(n+1)/2\}$$

$$\sum_{k=1}^{n} k^2 = n(n+1)(2n+1)/6$$

$$\sum_{k=1}^{n} n = n^2$$

and

$$\sum_{k=1}^{n} k = n(n+1)/2$$

Hence, the operation count for this stage is

$$n^3 - n^2(n+1) + n(n+1)(2n+1)/6 + n^2 - n(n+1)/2 = n^3/3 - n/3$$

operations. In the final stage, (3.12), there are $(n-1)$ values of i, for each of which there are $(n-i)$ values of j; for each of these, in turn, there is one multiplication operation. Thus, there are

$$1 + 2 + \cdots + (n+1) = n(n-1)/2$$

operations in this stage.

The operation count for the entire process is therefore

$$n(n+1)/2 + n^3/3 - n/3 + n(n-1)/2 = n^3/3 + n^2 - n/3$$

For large values of n the first term predominates; hence, we can say that the solution time is roughly proportional to the cube of the size of the system.

3.6 Tridiagonal systems

In the numerical solution of ordinary and partial differential equations, the need often arises to solve a linear system of the general form

$$
\begin{aligned}
a_{11}x_1 + a_{12}x_2 &\quad\quad\quad\quad\quad\quad\quad\quad\quad\quad = a_{1,n+1} \\
a_{21}x_1 + a_{22}x_2 + a_{23}x_3 &\quad\quad\quad\quad\quad\quad\quad\quad = a_{2,n+1} \\
a_{32}x_2 + a_{33}x_3 + a_{34}x_4 &\quad\quad\quad\quad\quad\quad = a_{3,n+1} \\
&\vdots \\
a_{i,i-1}x_{i-1} + a_{ii}x_i + a_{i,i+1}x_{i+1} &= a_{i,n+1} \\
&\vdots \\
a_{n,n-1}x_{n-1} + a_{nn}x_n &= a_{n,n+1}
\end{aligned}
\quad (3.15)
$$

in which there are no more than three unknowns appearing in each equation: all of the coefficients are zero except those on the leading diagonal and on each adjacent diagonal. Such a system is called *tridiagonal*, and may be solved by the full elimination algorithm described in the previous section. However, for such systems the elimination and back-substitution processes become considerably simpler and faster when advantage is taken of the fact that many or most of the coefficients are zero. This special version of the method of elimination is known as the *Thomas algorithm*.

For clarity, we rewrite (3.15) using the symbols a, b, c and d to denote the subdiagonal, diagonal and supradiagonal coefficients and the right-hand side, respectively, so that the ith equation becomes

$$a_i x_{i-1} + b_i x_i + c_i x_{i+1} = d_i \quad (3.16)$$

It will be seen that the first equation of the system, involving only x_1 and x_2, can be rearranged to express x_1 in terms of x_2, and hence be used to eliminate x_1 from the second equation. This, in turn, can be rearranged to express x_2 in terms of x_3, and be used to eliminate x_2 from the third equation. Continuing in this manner, we will eventually eliminate x_{n-1} from the nth and last equation. Since the nth equation only contained x_{n-1} and x_n in the first instance, it can now be solved for x_n. The remaining unknowns can then be found in the reverse order by back-substitution.

However, during the elimination process the coefficients being computed at each step become increasingly unwieldy. So, instead of computing them explicitly, we adopt the following tactic.

Suppose that, after x_{i-2} has been eliminated from the $(i-1)$th equation, that equation then reads

$$\beta_{i-1}x_{i-1} + \gamma_{i-1}x_i = \delta_{i-1} \quad (3.17)$$

where the values of β_{i-1}, γ_{i-1} and δ_{i-1} are yet to be determined. The next stage is the use of (3.17) to eliminate x_{i-1} from the ith equation, which is (3.16). From (3.17),

$$x_{i-1} = (\delta_{i-1} - \gamma_{i-1}x_i)/\beta_{i-1}$$

Inserting this value into (3.16), we obtain

$$(b_i - a_i\gamma_{i-1}/\beta_{i-1})x_i + c_i x_{i+1} = d_i - a_i\delta_{i-1}/\beta_{i-1} \qquad (3.18)$$

However, by analogy with (3.17), (3.16) should become

$$\beta_i x_i + \gamma_i x_{i+1} = \delta_i \qquad (3.19)$$

Comparing the coefficients in (3.18) and (3.19) we see that

$$\beta_i = b_i - a_i\gamma_{i-1}/\beta_{i-1} \qquad (3.20a)$$
$$\gamma_i = c_i \qquad (3.20b)$$
$$\delta_i = d_i - a_i\delta_{i-1}/\beta_{i-1} \qquad (3.20c)$$

Equations (3.20a) and (3.20c) can be used to compute the values of β and δ recursively; γ is unnecessary, and c may be retained.

The first equation of the system is already in the form (3.20), viz.

$$\beta_1 x_1 + c_1 x_2 = \delta_1$$

Thus, the elimination stage is

$$\beta_1 \leftarrow b_1;$$
$$\delta_1 \leftarrow d_1;$$
For $i = 2, 3, \ldots, n$:
$$\beta_i \leftarrow b_i - c_{i-1}/\beta_{i-1}$$
$$\delta_i \leftarrow d_i - \delta_{i-1}/\beta_{i-1}$$

The nth equation is

$$\beta_n x_n = \delta_n$$

and hence

$$x_n = \delta_n/\beta_n$$

The $(n - 1)$th equation is

$$\beta_{n-1}x_{n-1} + c_{n-1}x_n = \delta_{n-1}$$

or

$$x_{n-1} = (\delta_{n-1} - c_{n-1}x_n)/\beta_{n-1}$$

or, in general,

$$x_i = (\delta_i - c_i x_{i+1})/\beta_i, \qquad i = n-1, n-2, \ldots, 1$$

Finally, we note that the work of the elimination stage can be reduced somewhat by introducing a quantity

$$\varepsilon = a_i/\beta_{i-1}$$

since this division operation occurs in the calculation of both β_i and δ_i.

The complete Thomas algorithm is thus

$$\begin{aligned}
&\beta_1 \leftarrow b_1; \\
&\delta_1 \leftarrow d_1; \\
&\text{For } i = 2, 3, \ldots, n: \\
&\quad \varepsilon \leftarrow a_i/\beta_{i-1} \\
&\quad \beta_i \leftarrow b_i - \varepsilon c_{i-1} \\
&\quad \delta_i \leftarrow d_i - \varepsilon \delta_{i-1} \\
&x_n \leftarrow \delta_n/\beta_n. \\
&\text{For } i = n-1, n-2, \ldots, 1: \\
&\quad x_i \leftarrow (\delta_i - c_i x_{i+1})/\beta_i.
\end{aligned} \qquad (3.21)$$

Since we are only working with the non-zero elements – the elements on the three leading diagonals – the number of operations involved is clearly very much less than in the case of the general elimination routine. The student should verify that the number of multiplications and divisions is $5n - 4$; this fact makes the method superior to any other, direct or iterative, for tridiagonal systems.

Figure 3.2 shows the Fortran listing of a subroutine which implements the Thomas algorithm, together with a simple main program to test and demonstrate the subroutine. The equations solved in this demonstration are

$$\begin{aligned}
-4x_1 + x_2 &= 0.2 \\
x_1 - 4x_2 + x_3 &= 0.4 \\
x_2 - 4x_3 + x_4 &= 0.6 \\
x_3 - 4x_4 + x_5 &= 0.8 \\
x_4 - 4x_5 &= 1.0
\end{aligned}$$

Coefficients of this nature are relevant to the numerical solution of differential equations; the right-hand sides, however, are purely artificial. The accuracy of the solution should be checked by substitution.

```
c
        subroutine thomas (a,b,c,d,x,n)
c
        dimension a(20),b(20),c(20),d(20),x(20),beta(20),delta(20)
c
        beta(1)=b(1)
        delta(1)=d(1)
        do 10 i = 2, n
            epsilon=a(i)/beta(i-1)
            beta(i)=b(i)-epsilon*c(i-1)
   10       delta(i)=d(i)-epsilon*delta(i-1)
c
        x(n)=delta(n)/beta(n)
        do 20 i = n-1, 1, -1
   20       x(i)=(delta(i)-c(i)*x(i+1))/beta(i)
        return
        end
c
c       main program to demonstrate the use of subroutine thomas
c
        dimension a(20),b(20),c(20),d(20),x(20)
c
        n=5
        do 5 i=1,n
            a(i)=   1.
            b(i)=  -4.
            c(i)=   1.
    5       d(i)= i/5.
        call thomas (a,b,c,d,x,n)
        write (*,15) (i,x(i),i=1,5)
   15   format(' the solution is:'//('       x(',i2,',') = ',f7.4))
        stop
        end
```

Figure 3.2 The Thomas algorithm for tridiagonal systems.

The output from this program is:

```
          the solution is:

               x( 1) =  -.0992
               x( 2) =  -.1969
               x( 3) =  -.2885
               x( 4) =  -.3569
               x( 5) =  -.3392
```

3.7 Extensions to the Thomas algorithm

The technique known as the Thomas algorithm can be extended to the pentadiagonal system[*] which arises in the solution of fourth-order differential equations and elsewhere. The general equation in such a system can be written

$$a_i x_{i-2} + b_i x_{i-1} + c_i x_i + d_i x_{i+1} + e_i x_{i+2} = f_i, \qquad i = 1, 2, \ldots, n$$

By an argument similar to that used in the previous section, it can be shown that the solution of this system is given by

---

[*] This method has been taken from D. U. Von Rosenberg 1969. *Methods for the numerical solution of differential equations.* New York: American Elsevier.

$\delta_1 = d_1/c_1;$ $\quad \lambda_1 = e_1/c_1;$ $\quad \gamma_1 = f_1/c_1;$

$\mu = c_2 - b_2\delta_1;$ $\quad \delta_2 = (d_2 - b_2\lambda_1)/\mu;$
$\lambda_2 = e_2/\mu;$ $\quad \gamma_2 = (f_2 - b_2\gamma_1)/\mu;$

For $i = 3, 4, \ldots, n$:
$\quad \beta = b_i - a_i\delta_{i-2};$
$\quad \mu = c_i - \beta\delta_{i-1} - a_i\lambda_{i-2};$ (3.22)
$\quad \delta_i = (d_i - \beta\lambda_{i-1})/\mu;$
$\quad \lambda_i = e_i/\mu;$
$\quad \gamma_i = (f_i - \beta\gamma_{i-1} - a_i\gamma_{i-2})/\mu;$

$\beta = b_{n-1} - a_{n-1}\delta_{n-3};$ $\quad \mu = c_{n-1} - \beta\delta_{n-2} - a_{n-1}\lambda_{n-3};$
$\delta_{n-1} = (d_{n-1} - \beta\lambda_{n-2})/\mu;$ $\quad \gamma_{n-1} = (f_{n-1} - \beta\gamma_{n-2} - a_{n-1}\gamma_{n-3})/\mu;$

$\beta = b_n - a_n\delta_{n-2};$ $\quad \mu = c_n - \beta\delta_{n-1} - a_n\lambda_{n-2};$
$\gamma_n = (f_n - \beta\gamma_{n-1} - a_n\gamma_{n-2})/\mu$

$x_n = \gamma_n;$
$x_{n-1} = \gamma_{n-1} - \delta_{n-1}x_n;$

For $i = n - 2, n - 3, \ldots, 1$:
$\quad x_i = \gamma_i - \delta_i x_{i+1} - \lambda_i x_{i+2}$

The proof of the correctness of this algorithm, and its implementation in a program, are left as exercises for the student.

For banded matrices with a band width greater than five, which can be written in *block tridiagonal* form, the Thomas algorithm can be modified (*op. cit.*) to apply to the blocks which make up the coefficient matrix. Consider, for example, the following 12×12 system, in which a 't' has been used to denote any non-zero number, and zero elements in the coefficient matrix have been left blank:

$$\begin{bmatrix} t & t & t & t & & & & & & & & \\ t & t & t & t & t & & & & & & & \\ t & t & t & t & t & t & & & & & & \\ t & t & t & t & t & t & t & & & & & \\ & t & t & t & t & t & t & t & & & & \\ & & t & t & t & t & t & t & t & & & \\ & & & t & t & t & t & t & t & t & & \\ & & & & t & t & t & t & t & t & t & \\ & & & & & t & t & t & t & t & t & t \\ & & & & & & t & t & t & t & t & t \\ & & & & & & & t & t & t & t & t \\ & & & & & & & & t & t & t & t \end{bmatrix} \begin{bmatrix} x_1 \\ x_2 \\ x_3 \\ x_4 \\ x_5 \\ x_6 \\ x_7 \\ x_8 \\ x_9 \\ x_{10} \\ x_{11} \\ x_{12} \end{bmatrix} = \begin{bmatrix} t \\ t \\ t \\ t \\ t \\ t \\ t \\ t \\ t \\ t \\ t \\ t \end{bmatrix} \quad (3.23)$$

The only non-zero elements are those on the leading diagonal and on the three adjacent diagonals on each side of the leading diagonal. If the elements of the coefficient matrix are denoted by a_{ij}, then for each value of i from 1 to 12

$$a_{ij} \neq 0 \quad \text{for } |i - j| \leq 3$$
$$a_{ij} = 0 \quad \text{otherwise}$$

It is again possible to take advantage of the special structure of the matrix to reduce the effort below that required for a general matrix.

The coefficient matrix can be partitioned as shown; each block is (in this case) a 3×3 matrix. Denoting these blocks by upper case letters [corresponding to the lower case letters used in (3.16)], the system (3.23) can be written

$$\begin{bmatrix} B_1 & C_1 & & \\ A_2 & B_2 & C_2 & \\ & A_3 & B_3 & C_3 \\ & & A_4 & B_4 \end{bmatrix} \begin{bmatrix} X_1 \\ X_2 \\ X_3 \\ X_4 \end{bmatrix} = \begin{bmatrix} D_1 \\ D_2 \\ D_3 \\ D_4 \end{bmatrix} \qquad (3.24)$$

where X_i and D_i are 3×1 column vectors.

By an argument similar to that used in the previous section, it can be shown that the solution of a general $n \times n$ system [of which (3.24) is a 4×4 example] is given by

$$\Phi_1 = B_1^{-1}$$
$$\Psi_1 = D_1$$
For $i = 2, 3, \ldots, n$:
$$T = A_i \Phi_{i-1}$$
$$\Phi_i = (B_i - TC_{i-1})^{-1}$$
$$\Psi_i = D_i - T\Psi_{i-1}$$
$$X_n = \Phi_n \Psi_n$$
For $i = n - 1, n - 2, \ldots, 1$:
$$X_i = \Phi_i(\Psi_i - C_i X_{i+1})$$

Since B, C, T, Φ, etc., denote $m \times m$ matrices ($m = 3$ in the example), B^{-1}, etc., denote inversion. For small values of m (say $m = 3$) the simplest method of inversion is by determinants, i.e. by simply writing an explicit expression for each element in the inverse. For larger systems – that is, for systems in which the bandwidth is greater – it would be preferable to use an elimination algorithm for the $m \times m$ matrices. Even when this is necessary, (3.25) is faster than a general elimination routine for a banded system. Again, the derivation and implementation are left as exercises for the student.

3.8 Iterative methods for linear systems

When there is a need to solve very large linear systems, containing (say) 100, or perhaps 1000, or even more equations in as many unknowns, the method of elimination becomes very time-consuming (except for sparse systems, such as tridiagonal and pentadiagonal systems, for which special methods are available). Iterative methods may provide an alternative.

If the system (3.6) is rearranged so that the ith equation is explicit in x_i, we obtain

$$x_i = \frac{1}{a_{ii}}(a_{i,n+1} - a_{i1}x_1 - \cdots - a_{i,i-1}x_{i-1} - a_{i,i+1}x_{i+1} - \cdots - a_{in}x_n)$$

$$= \frac{1}{a_{ii}}\left(a_{i,n+1} - \sum_{\substack{j=1\\j\neq i}}^{n} a_{ij}x_j\right) \qquad (3.26)$$

The system is now in a form suitable for iteration, similar to that described in Chapter 2 for a single equation. Each equation of the type (3.26) is regarded as yielding a new estimate of one of the unknowns, and all such equations are used, in turn, to give new (and, hopefully, improved) values of the respective x. One iteration consists of the evaluation of n equations similar to (3.26).

It is now necessary to make a first estimate for *all* of the unknowns x_1, $x_2, \ldots, x_n$, a task which may be of some difficulty. If the system of equations has arisen from a problem in science or engineering, as would be expected here, then the nature of that problem might well provide a guide. Such a guide should always be sought, since the number of iterations to convergence depends on the quality of the initial estimate. (This is also true for single equations, but is more important here because the work per iteration is so much greater.) If there is no indication at all of the solution, then $x_1 = x_2 = \cdots = x_n = 0$ may be as good as any!

Denoting the vector of estimates after the kth iteration as $x_1^k, x_2^k, \ldots, x_n^k$, the $(k + 1)$th estimates are found from the iteration form of (3.26), viz.

$$x_i^{k+1} = \frac{1}{a_{ii}}\left(a_{i,n+1} - \sum_{\substack{j=1\\j\neq i}}^{n} a_{ij}x_j^k\right), \qquad i = 1, 2, \ldots, n \qquad (3.27)$$

This equation is used for each value of i in turn, and successive solution vectors are computed until convergence has been obtained. The procedure is called *Jacobi iteration* (it is sometimes also referred to as Richardson iteration).

Convergence is most easily measured in terms of the relative change in each value of x from one iteration to the next. If the quantities

$$d_i = \left|\frac{x_i^{k+1} - x_i^k}{x_i^{k+1}}\right|$$

are computed for each value of i, then convergence can be said to have been reached when each d_i is less than some specified small quantity. If the value of any of the quantities x_i is likely to be (or become) zero, this test will fail. A better test, although involving slightly more work, would then be to compute

$$d = \frac{\sum_{i=1}^{n} |x_i^{k+1} - x_i^k|}{\sum_{i=1}^{n} |x_i^{k+1}|}$$

and to require d to be less than some specified small quantity.

As in the case of a single equation, convergence is not always guaranteed. The *convergence criterion* for Jacobi iteration can be expressed in a number of ways. Probably the simplest is to say that the process (3.27) will converge if the system (3.6) is *diagonally dominant*, i.e. if, in each equation, the absolute value of the diagonal coefficient is greater than or equal to the sum of the absolute values of the off-diagonal elements, and in at least one equation is actually greater than that sum. In mathematical terms, (3.27) will converge if

$$|a_{ii}| \geq \sum_{\substack{j=1 \\ j \neq i}}^{n} |a_{ij}| \quad \text{for all } i$$

and

$$|a_{ii}| > \sum_{\substack{j=1 \\ j \neq i}}^{n} |a_{ij}| \quad \text{for at least one value of } i$$

This is a sufficient, but not always a necessary, condition for convergence. That is, if the coefficients satisfy this condition, then Jacobi iteration will definitely converge. If not, then convergence is not guaranteed; it may not, and probably will not, be achieved. In this case, it may be feasible to rearrange the equations to obtain diagonal dominance, although with large systems such rearrangement is not really practicable. The systems arising when finite difference methods are used to solve differential equations are normally diagonally dominant – indeed, the methods are normally chosen to ensure this – and iterative methods can then be used. When finite element methods are used, this is often not true, and direct methods must be employed even for very large systems.

To illustrate the use of Jacobi iteration, consider the system

$$\begin{aligned} 3x_1 + x_2 - x_3 &= 2 \\ x_1 - 4x_2 + x_3 &= -4 \\ x_1 - x_2 + 2x_3 &= 5 \end{aligned} \quad (3.28)$$

Note that the system satisfies the convergence criterion. It may be rewritten

ITERATIVE METHODS FOR LINEAR SYSTEMS

$$x_1 = (2 - x_2 + x_3)/3$$
$$x_2 = (4 + x_1 + x_3)/4 \qquad (3.29)$$
$$x_3 = (5 - x_1 + x_2)/2$$

Starting with the initial estimate $x_1 = x_2 = x_3 = 0$, for want of anything better, (3.29) yields (working to two decimal places)

$$x_1^1 = 0.67 \qquad x_2^1 = 1.00 \qquad x_3^1 = 2.50$$

for the first iteration, where the superscript identifies the number of the iteration. From this we obtain

$$x_1^2 = (2 - 1.00 + 2.50)/3 = 1.17$$
$$x_2^2 = (4 + 0.67 + 2.50)/4 = 1.79$$
$$x_3^2 = (5 - 0.67 + 1.00)/2 = 2.67$$

for the second iteration. The student should verify the next few iterations shown in Table 3.1.

As the iteration equations (3.27) make clear, when we come to calculate the $(k + 1)$th estimate of any of the unknowns $- x_i$, say $-$ we have already found the $(k + 1)$th values of $x_1, x_2, \ldots, x_{i-1}$, but we do not use them to find x_i. All of the unknowns used in the right-hand side of (3.27) are those found during the kth iteration.

It seems reasonable to suppose that it would be better to use in (3.27) any new values already known. It turns out that this is, indeed, the case. *Gauss–Seidel iteration* (also known as the Liebmann method) does just that. It replaces (3.27) by

$$x_i^{k+1} = \frac{1}{a_{ii}} \left(a_{i,n+1} - \sum_{j=1}^{i-1} a_{ij} x_j^{k+1} - \sum_{j=i+1}^{n} a_{ij} x_j^k \right), \quad i = 1, 2, \ldots, n \quad (3.30)$$

and, in effect, makes use at all times of the latest information available. The convergence criterion for Gauss–Seidel iteration is the same as that for Jacobi iteration, viz. that the system of equations should be diagonally dominant. Gauss–Seidel iteration converges about twice as quickly as Jacobi iteration, and would always be used in preference to it. Indeed, in a

Table 3.1 Solution of (3.28) by Jacobi iteration.

					k				
	0	1	2	3	4	5	6	7	8
x_1^k	0	0.67	1.17	0.96	0.95	1.02	1.00	0.99	1.00
x_2^k	0	1.00	1.79	1.96	1.94	1.99	2.00	2.00	2.00
x_3^k	0	2.50	2.67	2.81	3.00	3.00	2.98	3.00	3.00

Table 3.2 Solution of (3.28) by Gauss–Seidel iteration.

	k					
	0	1	2	3	4	5
x_1^k	0	0.67	1.19	0.97	1.01	1.00
x_2^k	0	1.17	1.99	1.97	2.00	2.00
x_3^k	0	2.75	2.90	3.00	3.00	3.00

computer program it is the easier and more normal method to use: the calculated value of a new estimate of x_i would overwrite the old estimate, and would therefore automatically be used in subsequent calculations involving x_i. Jacobi iteration requires more storage, because of the need to keep all the 'old' values until all the 'new' values have been found.

Table 3.2 shows the solution of (3.28) by Gauss–Seidel iteration. The increased rate of convergence over that in Table 3.1 is evident. This is typical of Gauss–Seidel iteration compared with Jacobi iteration.

A further improvement to the rate of convergence can sometimes be made. If x_i^k is added to the right-hand side of (3.30), and $(a_{ii}x_i^k)/a_{ii}$ is subtracted from it, we obtain

$$x_i^{k+1} = x_i^k + \frac{1}{a_{ii}} \left(a_{i,n+1} - \sum_{j=1}^{i-1} a_{ij} x_j^{k+1} - \sum_{j=i}^{n} a_{ij} x_j^k \right) \qquad (3.31)$$

(Note that the lower limit of the second summation has changed from $i + 1$ to i.) The second term on the right-hand side can be regarded as a correction which must be added to x_i^k to make it nearer to the correct value. As convergence is approached, the expression in parentheses tends to zero: it approaches, in fact, a rearrangement of the ith equation. If now this term is multiplied by a number ω, so that (3.31) becomes

$$x_i^{k+1} = x_i^k + \frac{\omega}{a_{ii}} \left(a_{i,n+1} - \sum_{j=1}^{i-1} a_{ij} x_j^{k+1} - \sum_{j=i}^{n} a_{ij} x_j^k \right) \qquad (3.32)$$

then in a sense we *overcorrect* or *undercorrect* x_i^k, depending on whether ω is, respectively, greater or less than unity. If ω is chosen properly, then (3.32) may converge faster than (3.31) does. This technique is known as the method of *successive over-relaxation* (SOR)*. The difficulty in its implemen-

---

* The term *relaxation* arises from the fact that much of the early work on methods for the iterative solution of systems of equations was related to the determination of forces and moments in structures. If incorrect values for these unknown quantities are used, then artificial restraining forces and moments are required at each node of the structure. As the correct solution is approached, these forces can be 'relaxed'. The method is known as *over*-relaxation even when, as sometimes happens (see below), the best value for ω is less than unity and under-relaxation is actually being used. The method is *successive*, because it is applied to each unknown successively, and in turn, rather than (as in another method which we shall not discuss) being applied to groups of, or even all of, the unknowns simultaneously. Like the other iterative methods for linear systems, it has another name which is sometimes used in the literature – the accelerated Liebmann method.

tation lies in finding the proper choice of ω. The optimum value can only be calculated analytically in special cases. These include the equations which arise in the numerical solution of Laplace's and Poisson's equations, where $1 < \omega < 2$; we shall discuss this matter further in Chapter 6. Otherwise, ω_{opt} – which depends only on the coefficient matrix – must be found by trial-and-error.

With systems of which the coefficient matrix is quite general (limited only by the requirement to be diagonally dominant), SOR is likely to offer little benefit over Gauss–Seidel iteration. The optimum value of ω, determined by numerical experiments for the particular system, will probably be close to unity. However, with strongly banded systems – especially those likely to be encountered in the finite difference solution of differential equations, SOR is very attractive.

Consider the system of N equations

$$-4x_1 + x_2 + x_4 = d_1$$
$$x_1 - 4x_2 + x_3 + x_5 = d_2$$
$$\cdot \quad \cdot \quad \cdot \quad \cdot \quad \cdot \quad \cdot \quad \cdot \quad \cdot \quad \cdot \quad \cdot$$
$$x_{i-3} + x_{i-1} - 4x_i + x_{i+1} + x_{i+3} = d_i \quad i = 3, 4, \ldots, N - 2$$
$$\cdot \quad \cdot \quad \cdot \quad \cdot \quad \cdot \quad \cdot \quad \cdot \quad \cdot \quad \cdot$$
$$x_{N-4} + x_{N-2} - 4x_{N-1} + x_N = d_{N-1}$$
$$x_{N-3} + x_{N-1} - 4x_N = d_N$$

This system has a structure similar to that which will be encountered when solving differential equations. Figure 3.3 shows the number of iterations required to solve this system (to a specified convergence criterion) for two values of N and over a range of values of ω. For $N = 10$ and $\omega = 1$ (which is Gauss–Seidel iteration), 44 iterations were required. At the optimum value of ω, which was found (by numerical experiment) to be 1.34, only 19 iterations were required. For $N = 25$ and $\omega = 1$, 159 iterations were

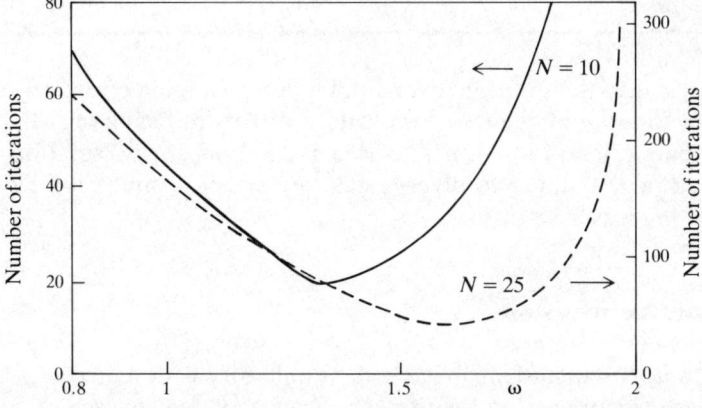

Figure 3.3 Number of iterations to convergence of SOR.

required, compared with 31 at $\omega_{opt} = 1.64$. Similar, although not identical, results can be expected for other strongly banded matrices. Students are encouraged to explore the use of SOR on matrices of their own invention.

It will be seen that SOR enables substantial savings to be achieved – if the optimum value of ω can be determined. It is also seen that – for a system of this type, at least – ω_{opt} increases with the size of the system. Finally, the figure illustrates – and it is generally true – that it is safer to underestimate ω, rather than overestimate it.

In addition to increasing the rate of convergence of systems which are already convergent, SOR can sometimes enable an iterative solution to be found for a system which is not convergent for Gauss–Seidel iteration. This is achieved by using $\omega < 1$, and therefore should properly be called SUR (successive under-relaxation), although this name is seldom used. Consider the system

$$1.81x_1 + 0.33x_2 + 0.88x_3 - 0.15x_4 + 0.83x_5 + 0.26x_6 = 3.29$$
$$0.69x_1 + 1.04x_2 - 0.93x_3 + 0.26x_4 - 0.36x_5 + 0.71x_6 = -3.97$$
$$0.92x_1 + 0.42x_2 + 2.27x_3 - 0.82x_4 + 0.90x_5 + 0.83x_6 = -5.30$$
$$0.64x_1 + 0.41x_2 - 0.57x_3 - 0.93x_4 + 0.99x_5 - 0.10x_6 = 3.90$$
$$-0.76x_1 - 0.31x_2 - 0.86x_3 - 0.08x_4 + 0.47x_5 - 0.47x_6 = -5.94$$
$$-0.02x_1 - 0.51x_2 - 0.02x_3 - 0.36x_4 - 0.76x_5 + 1.92x_6 = 1.29$$

This system is not diagonally dominant, and it was found that an attempt at a solution using Gauss–Seidel iteration rapidly diverged. However, the solution was capable of being found using 'SUR' with $\omega < 1$. The number of iterations required to obtain the solution, for various values of ω, is shown in Table 3.3.

Table 3.3 Use of under-relaxation to achieve convergence.

ω	0.40	0.45	0.50	0.55	0.60	0.65	0.70	0.75	0.80	0.85	0.90
N	26	23	20	18	16	14	13	12	17	35	858

For $\omega \geq 0.95$ the solution diverged. Under-relaxation effectively slows or dampens the rate of change of the values of x by reducing the effect of the term in parentheses in (3.32). This idea of damping down a solution which is diverging or is tending to diverge can be used in conjunction with many iterative processes.

3.9 Matrix inversion

It is assumed that students have some familiarity with matrix algebra, and are aware, in particular, of the significance of the inverse of a matrix.

Inversion of a matrix (A) requires the determination of the elements of a matrix (B) such that

$$(A)(B) = (I) \tag{3.33}$$

where (I) is the unit matrix. Equation (3.33) can be regarded as a set of n systems of equations, in any one of which one of the columns of (B) is the vector of unknowns, and the corresponding column of (I) is the known vector of right-hand side values. In the ith system, the unknowns are $b_{1i}, b_{2i}, \ldots, b_{ni}$, and the right-hand sides are all zero except for the ith, which is 1:

$$\begin{bmatrix} a_{11} & a_{12} & a_{13} & \cdots & & a_{1n} \\ a_{21} & a_{22} & a_{23} & \cdots & & a_{2n} \\ \vdots & \vdots & \vdots & \cdots & & \vdots \\ a_{i1} & a_{i2} & a_{i3} & \cdots & a_{ii} & \cdots & a_{in} \\ \vdots & \vdots & \vdots & \cdots & & \vdots \\ a_{n1} & a_{n2} & a_{n3} & \cdots & & a_{nn} \end{bmatrix} \begin{bmatrix} b_{1i} \\ b_{2i} \\ \vdots \\ b_{ii} \\ \vdots \\ b_{ni} \end{bmatrix} = \begin{bmatrix} 0 \\ 0 \\ \vdots \\ 1 \\ \vdots \\ 0 \end{bmatrix} \tag{3.34}$$

Thus, the inversion of (A) can be accomplished by solving n systems of equations like (3.34). Gaussian elimination is one way of doing this. It should be noted that since the coefficients in the n systems are the same – viz. the elements of (A) – the process of triangularization [i.e. the algorithm (3.9)] need only be performed once: the back-substitution (3.12) is then repeated for each of the n right-hand sides in turn.

How many operations are required to invert a matrix in this way? We have seen (Section 3.5) that normalization requires $n(n + 1)/2$ operations and the triangularization requires $(n^3 - n)/3$ operations. The total for (3.9) is therefore

$$n^3/3 + n^2/2 + n/6$$

Each back-substitution stage – of which there are n – requires $n(n - 1)/2$ operations. Thus, the total for inversion is

$$n^3/3 + n^2/2 + n/6 + n^2(n - 1)/2 = 5n^3/6 + n/6 \tag{3.35}$$

If the need exists to solve a number of systems of equations each of which has the same coefficient matrix but different vectors of right-hand side values, it might be thought that an efficient procedure would be to invert the coefficient matrix and then multiply the inverse into each right-hand side vector in turn. However, an operation count shows that it is more efficient to interrupt the inversion procedure after triangularization has been completed, and then perform the back-substitution process for each of the right-hand side vectors. From the foregoing calculations, the number of operations needed for the solution of m systems in this latter manner is

$$n^3/3 + n^2/2 + n/6 + mn(n - 1)/2 = n^3/3 + n^2(m + 1)/2 - n(3m - 1)/6 \tag{3.36}$$

On the other hand, each multiplication of a matrix by a vector involves n^2 operations (n multiplications to find each of the n elements of the solution vector). The count for inversion is (3.35). Thus, for m systems there is a total of

$$5n^3/6 + mn^2 + n/6 \qquad (3.37)$$

operations if complete inversion, followed by matrix multiplication, is performed. It is apparent that (3.36) is always less than (3.37).

3.10 The method of least squares

There are two topics involving the solution of linear systems that it is convenient to discuss at this stage. Each is concerned with curve-fitting or empirical analysis of data, i.e. the determination of a mathematical expression for the functional relationship between two variables based on experimental information. (We will not consider problems involving more than two variables.)

It is necessary to decide the most appropriate general form for the function, and then to determine the coefficients which cause the data to be satisfied as well as possible. We will consider two related methods, one suitable for functions in which the unknown coefficients appear linearly, and the other applicable to more general functions. The first is known as the *method of least squares*.

Suppose an experiment has been conducted to determine values of some quantity y as a function of some other quantity x. As a result of the experiment, n pairs of values of x and y have been measured. Suppose, further, that evidence exists to suggest that the relation between x and y is of the form

$$y(x) = a_0 + a_1 x + a_2 x^2 + \cdots + a_m x^m \qquad (3.38)$$

There are $(m + 1)$ unknown coefficients here $(a_0, a_1, \ldots, a_m)$ which appear linearly in (3.38). If the number of sets of data, n, is equal to $(m + 1)$, then we can construct $(m + 1)$ simultaneous equations and hence find $a_0, a_1, \ldots, a_m$. If n is less than $(m + 1)$, then there is insufficient information to enable the coefficients to be found, but if n is greater than $(m + 1)$ – and this is normally the case – then the problem is overspecified. It is not (normally) possible to choose values of $a_0, a_1, \ldots, a_m$ which will allow (3.38) to be satisfied exactly for each set of data values.

Instead, we are forced to settle for (3.38) being satisfied 'as well as possible'. Depending on what is meant by this phrase, the coefficients can then be determined.

The problem is illustrated in Figure 3.4, which shows a hypothetical set of data which might have been taken during the course of an experiment. There

THE METHOD OF LEAST SQUARES

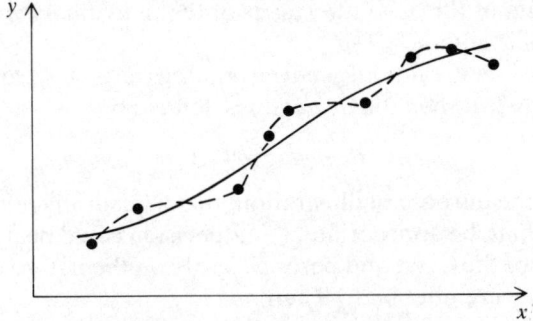

Figure 3.4 The problem of curve-fitting.

appears to be a functional relationship between the values of x and y: larger values of x generally have larger values of y associated with them, but the points are scattered about the curve defining the relationship. Of course, it would be possible to include sufficient coefficients in (3.38) – viz. n, the number of observations (nine, for the data of Fig. 3.4) – to allow a curve to be found which passes through every point. However, it is far more reasonable to suppose that the points include some experimental error, and that the solid line of Figure 3.4 is a better description of $y(x)$ than the broken line is.

This error could have occurred in the measurements of x, or of y, or of both. Frequently, we will have a situation in which one of the variables (say x) can be measured more accurately than the other. The errors are then attributed entirely to the y-values, as shown in Figure 3.5, and the problem is to choose the function $y(x)$ so that the individual errors $d_1, d_2, \ldots$ are somehow minimized. Since some of the ds are positive, and some negative, the usual approach is to minimize the sum of their squares: hence the name *least squares*. Other measures of the overall error could be used (for

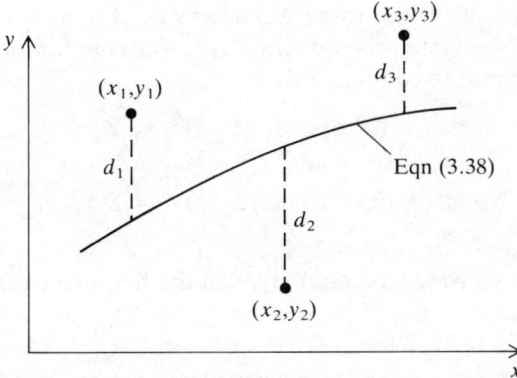

Figure 3.5 The definition of 'deviation' when the errors are attributed to y.

example, the sum of the absolute values of the individual errors), but least squares is the most common.

If y_i is the observed value of y corresponding to $x = x_i$, and $y(x_i)$ is the value given by (3.38), then the *deviation* is defined by

$$d_i = y_i - y(x_i)$$

Suppose, for the purpose of illustration, that a quadratic relation between x and y is thought to be appropriate. (This decision could be based partly on the appearance of the data, and perhaps partly on the nature of the process from which they were obtained.) Then

$$d_i = y_i - a_0 - a_1 x_i - a_2 x_i^2$$

and therefore the sum of the squares of the deviations is

$$\sum_{i=1}^{n} d_i^2 = \sum_{i=1}^{n} (y_i - a_0 - a_1 x_i - a_2 x_i^2)^2 = S \text{ (say)} \qquad (3.39)$$

If a_0, a_1 and a_2 are to be chosen so as to minimize (3.39), then

$$\frac{\partial S}{\partial a_0} = \frac{\partial S}{\partial a_1} = \frac{\partial S}{\partial a_2} = 0$$

Hence

$$\sum_{i=1}^{n} (y_i - a_0 - a_1 x_i - a_2 x_i^2) = 0$$

$$\sum_{i=1}^{n} (y_i - a_0 - a_1 x_i - a_2 x_i^2) x_i = 0$$

$$\sum_{i=1}^{n} (y_i - a_0 - a_1 x_i - a_2 x_i^2) x_i^2 = 0$$

These three equations in the three unknowns a_0, a_1 and a_2 can be readily solved, provided the system is not singular*. For computational purposes they can be written more conveniently as

$$na_0 + (\Sigma x_i) a_1 + (\Sigma x_i^2) a_2 = \Sigma y_i$$
$$\Sigma x_i a_0 + (\Sigma x_i^2) a_1 + (\Sigma x_i^3) a_2 = \Sigma x_i y_i \qquad (3.40)$$
$$\Sigma x_i^2 a_0 + (\Sigma x_i^3) a_1 + (\Sigma x_i^4) a_2 = \Sigma x_i^2 y_i$$

where all summations are for $i = 1, 2, \ldots, n$.

As an example, consider the data given in the first two columns of Table

---

* If the number of coefficients to be determined is large, then the system can become *ill-conditioned*: its determinant approaches zero. Alternative procedures must be employed.

Table 3.4 Least squares analysis of data.

x_i	y_i	$y(x_i)$	d_i
1.02	0.20	0.211	−0.011
5.34	3.34	5.267	−1.927
6.54	12.49	10.841	1.649
7.56	19.26	17.004	2.256
10.03	35.80	37.356	−1.556
13.18	69.04	74.453	−5.413
13.51	83.07	79.062	4.008
14.89	99.30	99.823	−0.523
17.57	151.18	146.989	4.191
18.92	171.50	174.173	−2.673

3.4. Since there are ten sets of values of x and y, $n = 10$. Computing the quantities required for (3.40), we obtain the system

$$10 \ a_0 + 108.56 a_1 + 1474.7 a_2 = 645.18$$
$$108.56 a_0 + 1474.7 \ a_1 + 22\,127.5 a_2 = 10\,016.2$$
$$1474.7 \ a_0 + 22\,127.5 \ a_1 + 352\,115 \ a_2 = 162\,564$$

and hence find that

$$a_0 = 2.446 \qquad a_1 = -2.833 \qquad a_2 = 0.6295$$

Thus, the 'best' relationship between x and y, in the least squares sense, is

$$y = 2.446 - 2.833x + 0.6295x^2 \qquad (3.41)$$

The values of y given by (3.41) are shown in the third column of Table 3.4, and the corresponding deviations are in the fourth column. It will be seen that (3.41) satisfies none of the data exactly but, because of the manner of its derivation, it represents a good compromise. Thus, the ten equations in three unknowns have been solved 'as well as possible'.

It is important to remember that this analysis was based on the assumption that all of the errors can be attributed to the y-values and that the x-values are exact. If all the errors can reasonably be attributed to the x-values, then it is only necessary to interchange x and y in the foregoing theory. However, situations can exist in which the error should be attributed equally to x and y. In such cases the deviations must be computed as indicated in Figure 3.6.

It is also important to remember that (3.41) is only the 'best' result for a least squares analysis. If we had sought to minimize the sum of the absolute values of the deviations, or some other measure of the deviations, a different equation may have been obtained.

The analyses of these, and other possibilities, are somewhat more complicated, and are not treated here. However, the student should be aware that the foregoing approach is not the only one which may be taken.

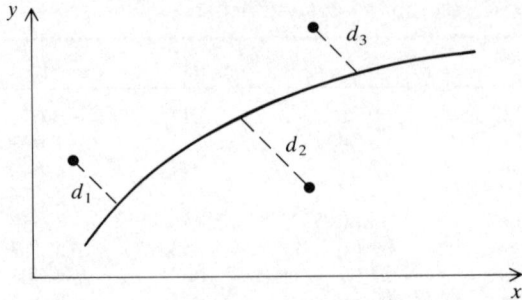

Figure 3.6 The definition of 'deviation' when the errors are attributed equally to x and y.

3.11 The method of differential correction

It can be seen that the least squares method of the previous section will only be applicable when the coefficients to be determined appear linearly in the functional relationship which is assumed to exist between x and y; otherwise the system (3.40) will not be linear. Often, however, the form of the data or the nature of the process from which they were obtained suggest a non-linear function. Consider, for example, the data of Table 3.5. The data have been plotted on a log–log graph in Figure 3.7, which strongly suggests that x and y are related by an expression of the form

$$y = ax^b \qquad (3.42)$$

Figure 3.7 further suggests that

$$a \approx 6 \quad \text{and} \quad b \approx -1 \qquad (3.43)$$

A least squares analysis of an equation of the form (3.42) leads to a non-linear system of equations which must be solved for a and b. While such systems can be solved (for example, by the methods to be described in the next section), the solution is often much more difficult than for a linear system of the same size.

As an alternative, we could seek to *linearize* the function. For example (3.42) can be written

$$Y = A + bX$$

where $X = \log x$, $Y = \log y$ and $A = \log a$. The values of X and Y can be

Table 3.5 Experimental data to be represented by $y = ax^b$.

x_i	1.45	2.05	2.25	2.40	3.30	3.95	4.10	5.12
y_i	4.53	3.46	3.19	3.00	2.31	1.98	1.91	1.58

THE METHOD OF DIFFERENTIAL CORRECTION

found from the data, and the least squares method can then be used to find A and b.

Often, however, this cannot be done: for example, it may be suggested that the data should be represented by a function of the form

$$y = ax^b + cx^d$$

Then y cannot be expressed as a linear combination of a, b, c and d. In such a case, the *method of differential correction* can be used. It relies on obtaining first estimates of the unknown coefficients, and leads to an iterative procedure involving the solution of a linear system for the *corrections* to the assumed values of the coefficients.

Suppose the data (x_i, y_i), $i = 1, 2, \ldots, n$, are to be represented by a non-linear function

$$y = f(x, a, b, c)$$

where a, b and c are the coefficients to be determined. Suppose further that first estimates a_0, b_0 and c_0 of these parameters have been found. Let α, β and γ be corrections to a_0, b_0 and c_0, respectively such that $(a_0 + \alpha)$, $(b_0 + \beta)$ and $(c_0 + \gamma)$ are better estimates of a, b and c. The *deviations* between the observed values of y and those calculated from the assumed function are given by

$$\begin{aligned} d_i &= y_i - f(x_i, a_0 + \alpha, b_0 + \beta, c_0 + \gamma) \\ &= y_i - \left\{ f(x_i, a_0, b_0, c_0) + \alpha \frac{\partial f}{\partial a_0} + \beta \frac{\partial f}{\partial b_0} + \gamma \frac{\partial f}{\partial c_0} + \cdots \right\} \end{aligned} \quad (3.44)$$

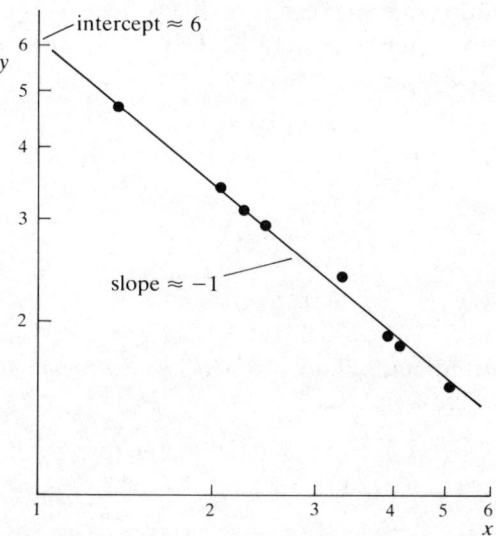

Figure 3.7 The data of Table 3.5 on a log–log graph.

where the partial derivatives are evaluated at (x_i, a_0, b_0, c_0). Since $a = a_0 + \alpha$, etc. (by definition of α), we may write

$$\frac{\partial f}{\partial a_0} = \frac{\partial f}{\partial a} = f_a \text{ (say)}$$

Equation (3.44) is now truncated after the first derivative terms, yielding

$$\begin{aligned} d_i &= y_i - f(x_i, a_0, b_0, c_0) - \alpha f_a - \beta f_b - \gamma f_c \\ &= r_i - \alpha f_a - \beta f_b - \gamma f_c \text{ (say)} \end{aligned}$$

where r_i is a quantity that can be computed from the data and the estimated values of a_0, b_0 and c_0. We now minimize Σd_i^2 by least squares, where

$$\Sigma d_i^2 = \Sigma (r_i - \alpha f_a - \beta f_b - \gamma f_c)^2 = S \text{ (say)}$$

The requirement that

$$\frac{\partial S}{\partial \alpha} = \frac{\partial S}{\partial \beta} = \frac{\partial S}{\partial \gamma} = 0$$

leads to equations for α, β and γ:

$$\begin{aligned} (\Sigma f_a^2)\alpha + (\Sigma f_a f_b)\beta + (\Sigma f_a f_c)\gamma &= \Sigma f_a r_i \\ (\Sigma f_a f_b)\alpha + (\Sigma f_b^2)\beta + (\Sigma f_b f_c)\gamma &= \Sigma f_b r_i \\ (\Sigma f_a f_c)\alpha + (\Sigma f_b f_c)\beta + (\Sigma f_c^2)\gamma &= \Sigma f_c r_i \end{aligned} \quad (3.45)$$

These three equations can be solved for α, β and γ. It is stressed that f_a, f_b and f_c, which denote the partial derivatives of $f(x, a, b, c)$ with respect to a, b and c, must be evaluated for each datum point using the estimates a_0, b_0 and c_0 when computing the coefficients in (3.45).

As an example, consider the data of Table 3.5, which we will try to represent by (3.42), using (3.2) as first estimates of a and b. Thus,

$$f(x, a, b) = ax^b$$

and

$$f_a \equiv \frac{\partial f}{\partial a} = x^b$$

$$f_b \equiv \frac{\partial f}{\partial b} = ax^b \ln x$$

The details of the computations are given in Table 3.6, which shows that the system to be solved for α and β is

$$\begin{aligned} 1.338\alpha + 6.022\beta &= 1.426 \\ 6.022\alpha + 33.238\beta &= 7.682 \end{aligned}$$

Hence,

$$\alpha = 0.131 \qquad \beta = 0.207$$

Table 3.6 Curve-fitting by the method of differential correction.

x_i	y_i	f_i	r_i	f_{ai}	f_{bi}	f_{ai}^2	f_{bi}^2	$f_{ai}f_{bi}$	$f_{ai}r_i$	$f_{bi}r_i$
1.45	4.53	4.138	0.392	0.690	1.538	0.476	2.364	1.060	0.270	0.603
2.05	3.46	2.927	0.533	0.488	2.101	0.238	4.414	1.025	0.260	1.120
2.25	3.19	2.667	0.523	0.444	2.162	0.198	4.676	0.961	0.233	1.132
2.40	3.00	2.500	0.500	0.417	2.189	0.174	4.790	0.912	0.208	1.094
3.30	2.31	1.818	0.492	0.303	2.171	0.092	4.712	0.658	0.149	1.068
3.95	1.98	1.519	1.461	0.253	2.087	0.064	4.354	0.528	0.117	0.962
4.10	1.91	1.463	0.447	0.244	2.065	0.059	4.264	0.504	1.109	0.922
5.12	1.58	1.172	0.408	0.195	1.914	0.038	3.663	0.374	0.080	0.781
						1.338	33.238	6.022	1.426	7.682

and

$$a = 6 + 0.131 = 6.131 \qquad b = -1 + 0.207 = 0.793$$

The relationship between x and y therefore becomes

$$y = 6.131x^{-0.793} \qquad (3.46)$$

Since the series in (3.44) was truncated, (3.46) will not be quite the 'best' relationship that can be obtained. The process can be repeated, using 6.131 and -0.793 as the current estimates of a and b, and new values of α and β obtained. The result is

$$\alpha = 0.073 \qquad \beta = -0.036$$

A further iteration yields

$$\alpha = 0.003 \qquad \beta = -0.001$$

Since α and β are now quite small, it is not worth performing another iteration. The final equation is

$$y = 6.207x^{-0.830}$$

3.12 Simple iteration for non-linear systems

A system of non-linear equations of the general form

$$\begin{aligned} f_1(\mathbf{x}) &= f_1(x_1, x_2, \ldots, x_n) = 0 \\ f_2(\mathbf{x}) &= f_2(x_1, x_2, \ldots, x_n) = 0 \\ &\cdots \cdots \cdots \cdots \cdots \\ f_n(\mathbf{x}) &= f_n(x_1, x_2, \ldots, x_n) = 0 \end{aligned} \qquad (3.47)$$

is usually solved by an iterative method, since it is not normally possible to use elimination. The simplest method is to rearrange (3.47) into the form

$$x_1 = F_1(x_1, x_2, \ldots, x_n) = F_1(\mathbf{x})$$
$$x_2 = F_2(x_1, x_2, \ldots, x_n) = F_2(\mathbf{x})$$
$$\cdots \cdots \cdots \cdots \cdots \cdots \cdots \cdots \quad (3.48)$$
$$x_n = F_n(x_1, x_2, \ldots, x_n) = F_n(\mathbf{x})$$

analogous to the rearrangement of the single equation (2.12) into (2.13).

The system (3.48) is now suitable for solution by simple iteration. An initial estimate of the solution vector must somehow be obtained – a task of considerably greater difficulty than in the case of a single equation. As suggested in Section 3.7 when discussing linear systems, the original problem which led to (3.47) will often provide a rough guide to the solution.

Successive estimates of the solution may now be computed iteratively from the iteration form of (3.48), viz.

$$x_1^{k+1} = F_1(x_1^k, x_2^k, \ldots, x_n^k) = F_1(\mathbf{x}^k)$$
$$x_2^{k+1} = F_2(x_1^k, x_2^k, \ldots, x_n^k) = F_2(\mathbf{x}^k)$$
$$\cdots \cdots \cdots \cdots \cdots \cdots \cdots \cdots \quad (3.49)$$
$$x_n^{k+1} = F_n(x_1^k, x_2^k, \ldots, x_n^k) = F_n(\mathbf{x}^k)$$

where the superscript k denotes the number of the iteration in which the value was computed. This method is a direct extension of (2.18) to systems, and is a form of Jacobi iteration, since the kth values are used throughout on the right-hand side of (3.49) until the $(k + 1)$th estimates of all n values of x have been found.

As in the case of (2.18), (3.49) does not always converge. Suppose that the solution is

$$\mathbf{S} \equiv (S_1, S_2, \ldots, S_n)$$

and that the error in the kth estimate is

$$\mathbf{e} \equiv (e_1^k, e_2^k, \ldots, e_n^k),$$

where

$$e_i^k = x_i^k - S_i, \quad i = 1, 2, \ldots, n$$

Then since $S_i = F_i(\mathbf{S})$, we may use the mean value theorem to write

$$x_i^{k+1} - S_i = F_i(\mathbf{x}^k) - F_i(\mathbf{S})$$
$$= \sum_{j=1}^{n} (x_i^k - S_i) \frac{\partial F_i\{\mathbf{S} + \xi_i^k(\mathbf{x}^k - \mathbf{S})\}}{\partial x_j} \quad (3.50)$$

where $0 \leq \xi \leq 1$.

Now suppose that $E^k = \max_i |x_i^k - S_i|$, i.e. that E^k is the largest element in the error vector at the kth iteration. Then (3.50) becomes

$$x_i^{k+1} - S_i \leq E^k \sum_{j=1}^{n} \left| \frac{\partial F_i}{\partial x_j} \right|, \quad i = 1, 2, \ldots, n \quad (3.51)$$

Since (3.51) is true for all i, it is true for that particular value, say I, which makes $|x_i^{k+1} - S_i|$ a maximum. Thus

$$E^{k+1} \leq E^k \sum_{j=1}^{n} \left|\frac{\partial F_I}{\partial x_j}\right| \tag{3.52}$$

Convergence requires that $E^{k+1} < E^k$, i.e. that the largest element (in absolute value) of the error vector at the $(k + 1)$th iteration should be less than the largest element in the kth error vector. Thus convergence is assured if

$$\sum_{j=1}^{n} \left|\frac{\partial F_I}{\partial x_j}\right| < 1 \tag{3.53}$$

Since I [i.e. the value of i in (3.51) which makes e_i^{k+1} a maximum] is not known – because S is not known – the condition (3.53) must be replaced by

$$\sum_{j=1}^{n} \left|\frac{\partial F_i}{\partial x_j}\right| < 1, \quad 1, 2, \ldots, n \tag{3.54}$$

That is, the condition must hold for *all* functions F in the system (3.48). It will then be sure to hold for the Ith function – whichever that happens to be. This will be sufficient to ensure convergence of (3.44).

Equation (3.54) is the extension to non-linear systems of the convergence criterion (2.22) for a single equation. It must be satisfied for the initial estimate $(\mathbf{x}^0)$ and for all values of $(\mathbf{x})$ between $(\mathbf{x}^0)$ and (S). In other words, it must be satisfied in some region around the solution, and the initial estimate must lie within that region – i.e. it must be 'sufficiently close' to the solution. As in Section 2.5, we cannot quantify what is meant by 'sufficiently close'.

Just as we found with linear systems, the rate of convergence can be improved by using the latest available estimates of the elements of $(\mathbf{S})$ at all times. Thus the ith equation of (3.49) becomes

$$x_i^{k+1} = F_i(x_1^{k+1}, x_2^{k+1}, \ldots, x_{i-1}^{k+1}, x_i^k, x_{i+1}^k, \ldots, x_n^k)$$

This is an extension of Gauss–Seidel iteration [cf. (3.30)], and also requires (3.54) to be satisfied for convergence.

As an example, consider the system

$$x_1^{1/2} + x_2^{1/2} - x_1 = -1$$
$$x_1^{1/2} + x_2^{1/2} - x_2 = 2$$

A suitable iteration form is

$$\begin{aligned} x_1^{k+1} &= (x_1^k)^{1/2} + (x_2^k)^{1/2} + 1 \\ x_2^{k+1} &= (x_1^{k+1})^{1/2} + (x_2^k)^{1/2} - 2 \end{aligned} \tag{3.55}$$

Starting with the initial estimate $(2, 2)$, the results of the first few iterations are found to be

| x_1 | 2 | 3.828 | 4.127 | 4.128 | 4.094 | 4.065 | 4.045 |
| x_2 | 2 | 1.371 | 1.201 | 1.128 | 1.086 | 1.058 | 1.040 |

and, if continued, the solution will be seen to converge to (4, 1). The student should verify that the system (3.55) satisfies the convergence criterion at $x = (2, 2)$, at solution $x = (4, 1)$, and at intermediate values, but that it does *not* satisfy this criterion at, for example, $x = (0, 0)$. Convergence would *not* be guaranteed with this initial estimate, and it can readily be verified that convergence to $S = (4, 1)$ is not, in fact, obtained.

3.13 Newton's method for non-linear systems

More rapid convergence to the solution of a non-linear system can be obtained by the use of Newton's method.

For clarity, we adopt a different notation, and for simplicity consider a system of only two equations, although the method applies to systems of any size. Suppose that the equations

$$\begin{aligned} f(x, y) &= 0 \\ g(x, y) &= 0 \end{aligned} \qquad (3.56)$$

have a solution (s, t). If (x_k, y_k) is an estimate of this solution, with an error (d_k, e_k), so that

$$s = x_k - d_k \quad \text{and} \quad t = y_k - e_k$$

then we may write

$$\begin{aligned} f(s, t) &= f(x_k - d_k, y_k - e_k) = 0 \\ g(s, t) &= g(x_k - d_k, y_k - e_k) = 0 \end{aligned}$$

Therefore, to first order

$$\begin{aligned} f(x_k, y_k) - d_k f_x(x_k, y_k) - e_k f_y(x_k, y_k) &= 0 \\ g(x_k, y_k) - d_k g_x(x_k, y_k) - e_k g_y(x_k, y_k) &= 0 \end{aligned} \qquad (3.57)$$

where the subscripts x and y on f and g denote partial differentiation.

Equation (3.57) is a *linear* system for the corrections d_k and e_k, which can be solved if the system is not singular, i.e. if

$$\begin{vmatrix} f_x & f_y \\ g_x & g_y \end{vmatrix} \neq 0$$

at (x_k, y_k) or at (s, t), or at any iterate in between.

At each iteration it is necessary to evaluate all of the functions (in this case two) and the derivatives of each function with respect to each variable (in this case a total of four derivatives), and then to solve a linear system (in this case, of two equations). It is therefore apparent that there can be a great deal of work per iteration.

As an example, consider the system of the previous section

$$f(x, y) = x^{1/2} + y^{1/2} - x + 1 = 0$$
$$g(x, y) = x^{1/2} + y^{1/2} - y - 2 = 0$$

for which

$$f_x(x, y) = \tfrac{1}{2}x^{-1/2} - 1$$
$$f_y(x, y) = \tfrac{1}{2}y^{-1/2}$$
$$g_x(x, y) = \tfrac{1}{2}x^{-1/2}$$
$$g_y(x, y) = \tfrac{1}{2}y^{-1/2} - 1$$

Starting with the initial estimate (2, 2), the results of the next two iterations are found to be (4.6213, 1.6213) and (4.0922, 1.0922). Convergence is clearly much faster than with simple iteration, when four iterations were required to reach roughly the same degree of accuracy. In general, the method is second order, and will almost always converge in fewer iterations than simple iteration does.

It is interesting to compare the work involved in the two methods. The most time-consuming operation (in this example) is finding the square root of a number (which is more efficiently done by using the library function SQRT than by raising the number to the power 0.5, and which takes perhaps 10 or 20 times as long as a multiplication operation). Three square roots must be found for each iteration of (3.55), whereas only two are needed for Newton's method*. The small amount of additional arithmetic in the latter would not prevent it from being considerably faster than simple iteration.

However, it is clear that for a more complex system, or for a larger system, this will not always be the case. Since it is necessary to solve a linear system (requiring of the order of $n^3/3$ operations) for the corrections at each iteration, the use of Newton's method may well be less efficient than the use of simple iteration, despite the smaller number of iterations needed.

In such cases a modification of Newton's method may be employed. Each equation of the system is regarded as though it were an equation for just one of the unknowns, and is 'solved' by using one iteration of Newton's method with the assumption that all the other unknowns are, in fact, known quantities.

Thus, to find (x_{k+1}, y_{k+1}) from (x_k, y_k) for the system (3.56), the equations

$$x_{k+1} = x_k - \frac{f(x_k, y_k)}{f_x(x_k, y_k)} \qquad (3.58a)$$

---

* Note that, for efficiency, $x^{1/2}$ and $y^{1/2}$ should be evaluated separately and stored *before* evaluation of the actual functions and their derivatives.

$$y_{k+1} = y_k - \frac{g(x_{k+1}, y_k)}{g_y(x_{k+1}, y_k)} \qquad (3.58b)$$

are used. Equation (3.58a) implies that $f(x, y) = 0$ is being solved for x alone; (3.58b) implies that $g(x, y) = 0$ is being solved for y alone. Note that x_{k+1} can be used in (3.58b), which should improve the rate of convergence.

The modified Newton's method requires less work per iteration than the full Newton's method does, since fewer function evaluations are needed, and it is not necessary to solve a linear system at each step. However, it will require more iterations (although fewer than simple iteration needs). For systems that are large or complex, or both, it will often require less overall work and is therefore the preferred method.

Convergence can often be enhanced by over-correcting, i.e. by replacing (3.58a) and (3.58b) by

$$x_{k+1} = x_k - \omega \frac{f(x_k, y_k)}{f_x(x_k, y_k)} \qquad (3.59a)$$

$$y_{k+1} = y_k - \omega \frac{g(x_k, y_k)}{g_y(x_k, y_k)} \qquad (3.59b)$$

where ω is a factor somewhat greater than unity.

This is, in fact, an extension of the method of successive over-relaxation [Eqn (3.32)] to non-linear systems. If (3.59a) is applied to the function

$$f_i(x) = a_{i,n+1} - \sum_{j=1}^{n} a_{ij} x_j$$

then (3.32) is obtained.

The system (3.55) was solved again, using the modified Newton's method. Using (2, 2) as the initial estimate, the method converged to (4 ± 0.001, 1 ± 0.001) in eight iterations at $\omega = 1$; only six iterations were required at $\omega = 1.2$; while ten iterations were needed at $\omega = 1.4$. As with successive over-relaxation of linear systems, it is safer to underestimate ω than to overestimate it.

Worked examples

1. Solve the system

$$4x - 2y + z = 25$$
$$x + y + z = 16$$
$$2x - y + 3z = 25$$

by Gaussian elimination, showing the full working of the algorithm (3.9) and (3.12).

Augmented matrix:
$$\begin{pmatrix} 4 & -2 & 1 & 25 \\ 1 & 1 & 1 & 16 \\ 2 & -1 & 3 & 25 \end{pmatrix}$$

$k = 1$; $a_{kj} \leftarrow a_{kj}/a_{kk}$:
$$\begin{pmatrix} 4 & -1/2 & 1/4 & 25/4 \\ 1 & 1 & 1 & 16 \\ 2 & -1 & 3 & 25 \end{pmatrix}$$

$i = 2, 3$; $a_{ij} \leftarrow a_{ij} - a_{ik}a_{kj}$:
$$\begin{pmatrix} 4 & -1/2 & 1/4 & 25/4 \\ 1 & 3/2 & 3/4 & 39/4 \\ 2 & 0 & 5/2 & 25/2 \end{pmatrix}$$

$k = 2$; $a_{kj} \leftarrow a_{kj}/a_{kk}$:
$$\begin{pmatrix} 4 & -1/2 & 1/4 & 25/4 \\ 1 & 3/2 & 1/2 & 13/2 \\ 2 & 0 & 5/2 & 25/2 \end{pmatrix}$$

$i = 3$; $a_{ij} \leftarrow a_{ij} - a_{ik}a_{kj}$:
(no change because $a_{32} = 0$)
$$\begin{pmatrix} 4 & -1/2 & 1/4 & 25/4 \\ 1 & 3/2 & 1/2 & 13/2 \\ 2 & 0 & 5/2 & 25/2 \end{pmatrix}$$

$k = 3$; $a_{kj} \leftarrow a_{kj}/a_{kk}$:
$$\begin{pmatrix} 4 & -1/2 & 1/4 & 25/4 \\ 1 & 3/2 & 1/2 & 13/2 \\ 2 & 0 & 5/2 & 5 \end{pmatrix}$$

Therefore $z = 5$
Therefore $y = 13/2 - (1/2)(5) = 4$
Therefore $x = 25/4 - (-1/2)(4) - (1/4)(5) = 7$

2. Solve the system
$$x + y/2 + z/3 = 1$$
$$x/2 + y/3 + z/4 = 0$$
$$x/3 + y/4 + z/5 = 0$$

(a) using exact arithmetic and (b) working only to *two significant* figures.

(a) Augmented matrix:
$$\begin{pmatrix} 1 & 1/2 & 1/3 & 1 \\ 1/2 & 1/3 & 1/4 & 0 \\ 1/3 & 1/4 & 1/5 & 0 \end{pmatrix}$$

After elimination:
$$\begin{pmatrix} 1 & 1/2 & 1/3 & 1 \\ 1/2 & 1/12 & 1 & -6 \\ 1/3 & 1/12 & 1/180 & 30 \end{pmatrix}$$

Therefore $z = 30$
$y = -6 - (1)(30) = -36$
$x = 1 - (1/2)(-36) - (1/3)(30) = 9$

(b) To two significant figures, the solution goes through the following steps:

$$\begin{pmatrix} 1.0 & 0.50 & 0.33 & 1.0 \\ 0.50 & 0.33 & 0.25 & 0 \\ 0.33 & 0.25 & 0.20 & 0 \end{pmatrix} \quad \begin{pmatrix} 1.0 & 0.50 & 0.33 & 1.0 \\ 0.50 & 0.08 & 0.09 & -0.5 \\ 0.33 & 0.08 & 0.09 & -0.33 \end{pmatrix}$$

$$\begin{pmatrix} 1.0 & 0.50 & 0.33 & 1.0 \\ 0.50 & 0.08 & 1.1 & -6.3 \\ 0.33 & 0.08 & 0.09 & -0.33 \end{pmatrix} \quad \begin{pmatrix} 1.0 & 0.50 & 0.33 & 1.0 \\ 0.50 & 0.08 & 1.1 & -6.3 \\ 0.33 & 0.08 & 0.002 & 0.17 \end{pmatrix}$$

$$\begin{pmatrix} 1.0 & 0.50 & 0.33 & 1.0 \\ 0.50 & 0.08 & 1.1 & -6.3 \\ 0.33 & 0.08 & 0.002 & 85 \end{pmatrix}$$

Therefore
$$z = 85$$
$$y = -6.3 - (1.1)(85) = -100$$
$$x = 1 - (0.5)(-100) - (0.33)(85) = 23$$

which is a disastrous result. Fortunately, most computers work to more than two significant figures (8, 10 or 15 are more typical values), but this example illustrates what can happen if insufficient precision is retained during the calculations.

Note: This coefficient matrix, in which each element a_{ij} is equal to $(i + j - 1)^{-1}$, is known as the *Hilbert* matrix. It is extremely ill-conditioned, i.e. it is nearly singular, and therefore provides a severe test of linear system solvers. The conditioning becomes worse as the order of the matrix increases.

3. Use the method of differential correction to find the coefficients in a correlation for the data in columns 1 and 2 of Table 3.7. The first step is to

Table 3.7 Calculations for Worked Example 3.

1 x_i	2 y_i	3 f_i	4 r_i	5 f_{ai}	6 f_{bi}
0.52	0.65	0.6427	0.0073	0.2571	1.2563
1.63	1.84	1.8193	0.0207	0.7277	2.7949
2.87	2.40	2.4770	−0.0770	0.9908	0.9713
4.05	2.21	2.2465	−0.0365	0.8986	−4.4423
5.23	1.29	1.2565	0.0335	0.5026	−11.3037
6.05	0.28	0.2908	−0.0108	0.1163	−15.0223
7.68	−1.66	−1.6075	−0.0525	−0.6430	−14.7047

WORKED EXAMPLES

determine the nature of a likely correlating function and estimates of the coefficients. If a graph of y against x is drawn, a function of the form

$$y = a \sin bx$$

is suggested, with initial estimates of the coefficients being

$$a = 2.5 \qquad b = 0.5$$

Thus, with $f(x) = 2.5 \sin 0.5x$ the values of $f_i = f(x_i)$, as given in column 3 of the table, can be computed. The residual $r_i = y_i - f_i$, given in column 3, can then be found. Columns 5 and 6 are calculated from

$$f_a = \sin bx \qquad f_b = ax \cos bx$$

for the respective values of x_i. Next, the quantities

$$\Sigma f_{ai}^2 = 3.0644 \qquad \Sigma f_{ai} r_i = -0.0428$$
$$\Sigma f_{bi}^2 = 599.7368 \qquad \Sigma f_{bi} r_i = 0.7101$$
$$\Sigma f_{ai} f_{bi} = 1.3538$$

can be obtained, leading to the equations

$$3.0644\alpha + 1.3538\beta = -0.0428$$
$$1.3538\alpha + 599.7368\beta = 0.7101$$

for α and β, of which the solutions are

$$\alpha = -0.0145 \qquad \beta = 0.0012$$

The next estimates of a and b are therefore

$$a = 2.4855 \qquad b = 0.5012$$

4. Solve the non-linear system

$$f(x, y) = x + y^2 - 0.1 = 0$$
$$g(x, y) = x^2 + y - 0.2 = 0$$

(a) by Jacobi iteration, (b) by Gauss–Seidel iteration, (c) by the full Newton's method and (d) by the simplified Newton's method.

To obtain a first estimate of the solution, we note that both x and y appear to be small – much less than unity – so that x^2 and y^2 will be even smaller. Therefore

$$x \approx 0.1 \qquad y \approx 0.2$$

should be a reasonable first approximation to the solution.

(a) To use iteration (either Jacobi or Gauss–Seidel), an obvious rearrangement of the equations is

$$x = 0.1 - y^2 = F(x, y)$$
$$y = 0.2 - x^2 = G(x, y)$$

For both iterative methods, the convergence criterion is

$$|F_x| + |F_y| < 1$$
$$|G_x| + |G_y| < 1$$

Using the initial estimates derived above, we calculate

$$F_x = 0 \quad F_y = -2y = -0.4 \quad \therefore |F_x| + |F_y| = 0.4$$
$$G_x = -2x = -0.2 \quad G_y = 0 \quad \therefore |G_x| + |G_y| = 0.2$$

The convergence criterion is therefore satisfied at the initial estimate. Strictly, it should be applied at the solution – but since we do not know what that is, the initial estimate will have to suffice.

Jacobi iteration uses

$$x_{n+1} = 0.1 - y_n^2$$
$$y_{n+1} = 0.2 - x_n^2$$

and yields

| x | 0.1 | 0.06 | 0.0639 | 0.0614 | 0.0616 | 0.0615 |
| y | 0.2 | 0.19 | 0.1964 | 0.1959 | 0.1962 | 0.1962 |

The solution has converged to four decimal places after five iterations.

(b) Gauss–Seidel iteration uses

$$x_{n+1} = 0.1 - y_n^2$$
$$y_{n+1} = 0.2 - x_{n+1}^2$$

and yields

| x | 0.1 | 0.06 | 0.0614 | 0.0615 |
| y | 0.2 | 0.1964 | 0.1962 | 0.1962 |

The solution has now converged to the same degree of accuracy as in (a) in three iterations, i.e. about twice as fast as Jacobi iteration.

(c) For Newton's method, we return to the equations in their original form, and compute

$$f_x = 1 \quad f_y = 2y$$
$$g_x = 2x \quad g_y = 1$$

Equations (3.57) then become

$$\Delta x + 2y \Delta y = 0.1 - x - y^2$$
$$2x \Delta x + \Delta y = 0.2 - x^2 - y$$

Inserting $x = 0.1$ and $y = 0.2$, we obtain
$$\Delta x + 0.4\Delta y = -0.04$$
$$0.2\Delta x + \Delta y = -0.01$$

of which the solution is
$$\Delta x = -0.0391 \qquad \Delta y = -0.0022$$

so that the next estimate of the solution is
$$x = 0.0609 \qquad y = 0.1978$$

Using these values of x and y, (3.57) now become
$$\Delta x + 0.3956\Delta y = -0.000\,024\,84$$
$$0.1218\Delta x + \Delta y = -0.001\,508\,81$$

of which the solution is
$$\Delta x = 0.0006 \qquad \Delta y = -0.0016$$

so that the next estimate of the solution is
$$x = 0.0615 \qquad y = 0.1962$$

Newton's method has converged in two iterations – even faster than Gauss–Seidel iteration (in this example), although there is rather more work per iteration.

(d) The equations for the simplified Newton's method are (3.58). For the given system, they become

$$x_{n+1} = x_n - \frac{x_n + y_n^2 - 0.1}{1} = 0.1 - y_n^2$$

$$y_{n+1} = y_n - \frac{x_{n+1}^2 + y_n - 0.2}{1} = 0.2 - x_{n+1}^2$$

As it happens, in this example the simplified Newton's method turns out to be the same as Gauss–Seidel iteration. This is a coincidence, and does not usually occur.

Problems

1. (a) Construct a 3×3 matrix A of elements selected at random within the range $(-1, 1)$ and a vector B of length 3.

(b) Solve the system $Ax = B$ by hand, using Gaussian elimination and following the pattern of steps set out in Worked Example 1. Verify your solution by substitution.

2. Extend the Gaussian elimination method to incorporate the idea of *row interchange*, mentioned in the last paragraph of Section 3.4. The purpose of this is to ensure that the diagonal element used at each stage (i.e. each value of a_{kk}) is the largest available in absolute value. This may be achieved by examining whether, for each value of k, and for any row i in the range $k < i \leqslant n$, the following condition is satisfied:

$$|a_{ik}| > |a_{kk}|$$

If so, rows i and k are interchanged. Since this is equivalent merely to writing the remaining equations in a different order, the system and its solution are unaffected.

Repeat Problem 1 using the extended algorithm.

3. Use a computer program based on Figure 3.1 to perform the following.

(a) Generate a square matrix A of specified size (no more than about 10×10, to prevent the computations becoming too time-consuming) and a column vector B of the same size.

(b) Solve the system $Ax = B$ using the elimination algorithm (3.7), (3.12). (Write the algorithm as a subroutine so that it may be extracted and used independently elsewhere when needed.) Include in your program a section to terminate the calculations if a diagonal element is too small – say, less (in absolute value) than 10^{-4}. Verify your program and solution by substitution.

(c) Extend the program to incorporate row interchange.

4. Use your computer program to solve a system in which the coefficients form the Hilbert matrix (see Worked Example 2). How large a system can be handled before the answers become unacceptable?

5. For each of the following sets of data, use the method of least squares to find a polynomial relationship of the form (3.38) between x and y. Graph each data set on plain paper to help choose the order of the polynomial and initial values for the coefficients.

(a) x 0.74   1.37   2.51   3.11   3.89
    y 6.35   2.16   1.12   0.62   0.52

(b) x 0.37   0.91   1.42   1.89   2.41   2.88
    y 3.18   2.14   1.68   1.44   1.27   1.18

(c) x 0.51   0.88   1.23   1.74   2.02
    y 3.20   3.90   4.52   5.99   6.85

(d) x 1.47   3.94   6.82   11.14   16.02   20.51
    y 0.41   1.21   1.92   2.96   3.46   3.42

(e) x 0.20   1.10   2.20   2.90   3.80   5.00   6.10
    y 3.62   0.51 -1.20 -1.88 -2.61 -3.49 -4.27

Calculate the root mean square (r.m.s.) value of the deviations between the given values of y and the values from your functions. Investigate how the r.m.s. deviation varies for different orders of polynomial.

6. For each of the sets of data in Problem 5 use the method of differential correction to find a two-parameter functional relationship

$$y = f(x, a, b)$$

between x and y. Graph each data set on plain, semi-log and log–log paper to help choose the form of the function and initial values for the parameters a and b.

Calculate the r.m.s. value of the deviations between the given values of y and the values from your functions. Investigate how the r.m.s. deviation varies for different assumed functions.

7. Repeat Problem 6 with a three-parameter function

$$y = f(x, a, b)$$

Calculate the r.m.s. value of the deviations between the given values of y and the values from your functions. Investigate how the r.m.s. deviation varies for different assumed functions. Investigate also whether this quantity is smaller when a two-parameter function $y = f(x, a, b)$ or a three-parameter function $y = f(x, a, b, c)$ is used.

Additional problems can be generated by the student. To construct a problem requiring the solution of a linear or non-linear system it is necessary only to invent the 'left-hand sides', choose a solution and then compute the 'right-hand sides'. For practice, or to test programs for the methods of least squares and differential correction, sets of data values can be constructed using a calculator, or may be found in mathematical tables, tables of property values, etc.

4

Interpolation, differentiation and integration

4.1 Introduction

In this chapter we introduce the concept of finite difference operators, and use these operators to develop formulae for interpolation, differentiation and integration of tabular data. The data may literally be set out in a table, as with experimental results, or they may be generated during a computation, for example during the solution of a differential equation.

If two variables x and y are functionally related, then the function may be known explicitly. For example, if

$$y = x^2$$

then it is possible, analytically, to determine y for any value of x, or to find dy/dx or $\int y \, dx$ over a given range of values of x.

However, if we are given the data in Table 4.1, then although it seems reasonable to suppose that some relationship which could be written $y = f(x)$ exists, we do not know what that relationship is. Accordingly, it is not possible to find the value of y corresponding to a value of x which does not appear in the table, nor can we find – for any value of x – the derivative or integral of the function.

The need to perform such operations on tabular data arises in two circumstances. First, the data may have been generated by an experiment. For example, suppose that in Table 4.1, x denotes time in seconds and y denotes the speed of a vehicle in metres per second. Then we must *interpolate* if we wish to know the speed of the vehicle at, say, $x = 3.4$ s after the start of motion. To find the acceleration at any time, we must *differentiate* the speed, i.e. compute dy/dx at that time. Also, to find the distance

Table 4.1 Hypothetical experimental data.

x	0	1	2	3	4	5	6	7	8
y	0	0.21	0.51	0.96	1.66	2.68	4.10	5.98	8.40

INTRODUCTION

travelled in a given period, we need to *integrate* the speed, i.e. calculate $\int y \, dx$ over that period. These quantities must be calculated *without knowing the analytical relationship between x and y*.

Recalling the curve-fitting techniques discussed in Chapter 3, we realize that we may adopt the approach of finding some functional relationship which fits the data 'as well as possible', even if it is not the 'true' function, and then perform the necessary operations – interpolation, differentiation or integration – on this function. However, to do this without excessive labour we must restrict ourselves to functions which are relatively simple and for which the unknown coefficients can be easily calculated from the data. It has been found that polynomial approximation satisfies these requirements well, especially if the data are equally spaced: i.e. if the differences between successive values of the independent variable – x in the above example – are equal.

The second situation in which we have to handle tabular data is in the numerical solution of differential equations. As we shall see in the next chapter, this involves generating a sequence of values, often (but not necessarily) at equal intervals of the independent variable. These values satisfy a given expression for the derivative of one variable with respect to the other, but the functional relationship between the variables themselves is not known.

The development of techniques for numerical interpolation, differentiation and integration is greatly helped by the use of what are known as *finite difference operators*.

The student will already be familiar with the concept of a differential operator. For example, the differential equation

$$\frac{d^2y}{dx^2} - 3\frac{dy}{dx} + 2y = 0$$

is sometimes written

$$(D^2 - 3D + 2)y = 0$$

where D denotes the operator d/dx. Here, D^2 does not indicate D multiplied by D, nor is the expression in parentheses to be multiplied by y. The symbols D, D^2 and $(D^2 - 3D + 2)$ denote *operators* which perform certain operations (but not multiplication!) upon y. We now introduce other symbols, drawn from the Roman and Greek alphabets, and known as finite difference operators, to denote other mathematical operations on data which are contained in a table of values.

The term 'finite difference' refers to the fact that in a table such as Table 4.1 the difference between any pair of successive values of either variable is finite, and not infinitesimal. We will generally be dealing with quantities whose values differ by small amounts, but we will not – as we do, for example, in differentiation – proceed to the limit in which these differences tend to zero.

4.2 Finite difference operators

We have recalled that the operator D is used to denote the operation d/dx:

$$Df(x) \equiv \frac{d}{dx} f(x) \tag{4.1}$$

Although D is *not* a finite difference operator, we shall make use of this notation, and we now seek to express the operation of differentiation – which *cannot* be performed since we do not know $f(x)$ – in terms of other operations which *can* be performed.

The notation for the various *finite difference* operators will now be defined. We imagine the operations they denote to be performed on some function $f(x)$ of x, and limit ourselves to *equally spaced* values of the independent variable x. The interval between successive x values is denoted by h. It must be remembered that the function $f(x)$ may not, and almost certainly will not, be known analytically. We are simply assuming that we have a variable y whose value is related somehow to the value of another variable x, a relationship which we describe with the notation $f(x)$.

The first operator listed below is the *shift operator* E. Its definition is given in (4.2), and its effect is to denote, in symbols, the value of the function with the argument '$x + h$', rather than 'x'. It is important to remember that E is an operator: it does not have a value. But $Ef(x)$ is a quantity with a value: namely, the value of $f(x + h)$. This is true for all of the finite difference operators.

Shift operator

$$Ef(x) = f(x + h) \tag{4.2}$$

Forward difference operator

$$\Delta f(x) = f(x + h) - f(x) \tag{4.3}$$

Backward difference operator

$$\nabla f(x) = f(x) - f(x - h) \tag{4.4}$$

Central difference operator

$$\delta f(x) = f(x + h/2) - f(x - h/2) \tag{4.5}$$

Average operator

$$\mu f(x) = \{f(x + h/2) + f(x - h/2)\}/2 \tag{4.6}$$

Integral operator

$$If(x) = \int_x^{x+h} f(x)\, dx \tag{4.7}$$

The last of these, I, is (like D) not a finite difference operator: it involves infinitesimal quantities, rather than finite quantities. We cannot compute the effects of using D and I, because we are only working with tabular data. It is our purpose here to see how we can use the other operators, whose effects we *can* determine, to obtain approximate values for $Df(x)$ and $If(x)$.

The expressions (4.3), (4.4) and (4.5) are called the *first forward difference*, *first backward difference* and *first central difference*, respectively, of $f(x)$.

The significance of some of these operators can be illustrated with reference to Table 4.1, in which $h = 1$. Thus,

$$f(3) = 0.96$$
$$Ef(3) = f(4) = 1.66$$
$$\Delta f(3) = f(4) - f(3) = 0.70$$
$$\nabla f(3) = f(3) - f(2) = 0.45$$

The results of the operators δ and μ similarly cannot be evaluated yet, because they involve what may be called 'half-interval' values: quantities whose values depend on the values of x halfway along an interval, and which therefore do not exist – or at least, are not known. For example,

$$\delta f(3) = f(3.5) - f(2.5)$$

and

$$\mu f(3) = \{f(3.5) + f(2.5)\}/2$$

but the values of $f(2.5)$ and $f(3.5)$ are not known. We shall shortly see what can be done about this.

It is possible to combine operators into compound expressions. As the following example illustrates, it is understood that the operations are performed from *right* to *left*, so that the operator written next to the function is performed first:

$$\begin{aligned} E\Delta f(x) &= E\{\Delta f(x)\} \\ &= E\{f(x+h) - f(x)\} \\ &= Ef(x+h) - Ef(x) \\ &= f(x+2h) - f(x+h) \end{aligned} \tag{4.8}$$

The last step follows from the fact that if

$$Ef(x) = f(x+h)$$

then

$$Ef(x + h) = f\{(x + h) + h\} = f(x + 2h)$$

Consider now

$$\begin{aligned} \Delta E f(x) &= \Delta\{Ef(x)\} \\ &= \Delta f(x + h) \\ &= f(x + 2h) - f(x + h) \end{aligned} \quad (4.9)$$

A comparison of (4.8) and (4.9) shows that the operators E and Δ are *commutative*. This is true for all of the finite difference operators.

If an operator is repeated, for example

$$EEf(x) = Ef(x + h) = f(x + 2h)$$

then we can use the notation E^2. This means* that the operation is to be performed twice. Thus

$$\begin{aligned} \Delta^2 f(x) &= \Delta\{\Delta f(x)\} \\ &= \Delta\{f(x + h) - f(x)\} \\ &= f(x + 2h) - 2f(x + h) + f(x) \end{aligned} \quad (4.10)$$

The expression (4.10) is referred to as the *second forward difference* of $f(x)$.

The concept of repeated operations is extended to include

$$E^n f(x) = EEE \cdots Ef(x) = f(x + nh) \quad (4.11)$$

Furthermore, we allow negative values of n:

$$E^{-1}f(x) = f(x - h)$$
$$E^{-n}f(x) = f(x - nh)$$

and fractional values:

$$E^{1/2}f(x) = f(x + h/2)$$

and even the value zero:

$$E^0 f(x) = f(x)$$

In this last case, E^0 is usually denoted by 1, which here stands for the *identity operator* or 'do nothing' operator.

E^{-1} is an *inverse* operator. Its effect is to reverse whatever was done by E:

$$\begin{aligned} E^{-1}Ef(x) &= E^{-1}f(x + h) \\ &= f(x) \end{aligned}$$

---

* Since E is not a variable with a value, E^2 cannot be intended to mean $E \times E$ – this has no meaning and there is therefore no ambiguity. The usage is analogous to that of D^2 to denote d^2/dx^2.

FINITE DIFFERENCES OPERATORS

The identity operator allows us to introduce further inverse operators, by the definitions

$$\Delta^{-1}\Delta \equiv 1$$
$$\nabla^{-1}\nabla \equiv 1$$
$$\delta^{-1}\delta \equiv 1$$

We can now derive some relationship between operators which will be useful in later work:

$$\begin{aligned}\Delta f(x) &= f(x+h) - f(x) \\ &= \mathrm{E}f(x) - f(x) \\ &= (\mathrm{E}-1)f(x)\end{aligned}$$

or, for brevity

$$\Delta \equiv \mathrm{E} - 1 \tag{4.12}$$

It should be continually stressed that relationships such as (4.12) are not equations, but statements of the equivalence of operators; (4.12) should be read as 'the effect of the operator Δ is the same as the effect of the compound operator $(\mathrm{E} - 1)$'.

The student should now verify that

$$\nabla \equiv 1 - \mathrm{E}^{-1} \tag{4.13}$$
$$\delta \equiv \mathrm{E}^{1/2} - \mathrm{E}^{-1/2} \tag{4.14}$$
$$\mu \equiv (\mathrm{E}^{1/2} + \mathrm{E}^{-1/2})/2 \tag{4.15}$$
$$\delta^2 \equiv \Delta - \nabla \tag{4.16}$$

We can incorporate the derivative operator D by an interesting extension of the notation:

$$\begin{aligned}\mathrm{E}f(x) &= f(x+h) \\ &= f(x) + hf'(x) + \frac{h^2}{2!}f''(x) + \frac{h^3}{3!}f'''(x) + \cdots \\ &= f(x) + h\mathrm{D}f(x) + \frac{h^2}{2!}\mathrm{D}^2 f(x) + \frac{h^3}{3!}\mathrm{D}^3 f(x) + \cdots \\ &= \left(1 + h\mathrm{D} + \frac{h^2\mathrm{D}^2}{2!} + \frac{h^3\mathrm{D}^3}{3!} + \cdots\right)f(x) \\ &= e^{h\mathrm{D}}f(x)\end{aligned} \tag{4.17}$$

Thus,

$$\mathrm{E} \equiv e^{h\mathrm{D}} \tag{4.18}$$

This statement is rather startling at first sight. What can be meant by raising a number ($e = 2.718\ldots$) to the power $h\mathrm{D}$, when D is an operator? It should

Table 4.2 The relationships between the finite difference operators.

	E	Δ	∇	δ	μ	hD
E		$1 + \Delta$	$(1 - \nabla)^{-1}$	$1 + \frac{1}{2}\delta^2 + \delta(1 + \frac{1}{4}\delta^2)^{1/2}$	$2\mu^2 + 2\mu(\mu^2 - 1)^{1/2} - 1$	e^{hD}
Δ	$E - 1$		$\nabla(1 - \nabla)^{-1}$	$\delta(1 + \frac{1}{4}\delta^2)^{1/2} + \frac{1}{2}\delta^2$	$2\mu^2 + 2\mu(\mu^2 - 1)^{1/2} - 2$	$e^{hD} - 1$
∇	$1 - E^{-1}$	$\Delta(1 + \Delta)^{-1}$		$\delta(1 + \frac{1}{4}\delta^2)^{1/2} - \frac{1}{2}\delta^2$	$-2\mu^2 + 2\mu(\mu^2 - 1)^{1/2} - 2$	$1 - e^{-hD}$
δ	$E^{1/2} - E^{-1/2}$	$\Delta(1 + \Delta)^{-1/2}$	$\nabla(1 - \nabla)^{-1/2}$		$2(\mu^2 - 1)^{1/2}$	$2\sinh\frac{1}{2}hD$
μ	$\frac{1}{2}(E^{1/2} + E^{-1/2})$	$(1 + \frac{1}{2}\Delta)(1 + \Delta)^{-1/2}$	$(1 - \frac{1}{2}\nabla)(1 - \nabla)^{-1/2}$	$(1 + \frac{1}{4}\delta^2)^{1/2}$		$\cosh\frac{1}{2}hD$
hD	$\ln E$	$\ln(1 + \Delta)$	$-\ln(1 - \nabla)$	$2\sinh^{-1}\frac{1}{2}\delta$	$2\cosh^{-1}\mu$	hD

be realized that this is merely notational convenience. The *operator* e^{hD} is equivalent to the compound operator appearing in (4.17). Notice that D appears here accompanied by the increment h. We will see that this is often the case.

In a similar vein, we may utilize the facts that

$$\sinh \theta = (e^\theta - e^{-\theta})/2$$

and

$$\cosh \theta = (e^\theta + e^{-\theta})/2$$

together with (4.14) and (4.15) to write

$$\delta \equiv 2 \sinh hD/2 \qquad (4.19)$$

$$\mu \equiv \cosh hD/2 \qquad (4.20)$$

The student should now verify the relationships shown in Table 4.2, in which each of the operators E, Δ, ∇, δ, μ and hD are expressed in terms of each of the other operators alone. The integral operator I does not fit conveniently into this table, and we will defer its further consideration until Section 4.12.

4.3 Difference tables

Before the advent of high-speed digital computers, the numerical treatment of tabular data was often helped by the construction of *difference tables*. Although the need for such tables as part of the actual calculation process is now virtually non-existent, they still provide a useful aid to learning and understanding the concepts to follow. They also assist us to choose the most appropriate polynomial approximation to use in interpolation or numerical differentiation and integration.

Table 4.3 is an example of a *forward difference table*, and has been

Table 4.3 A forward difference table based on Table 4.1.

x	$f(x)$	$\Delta f(x)$	$\Delta^2 f(x)$	$\Delta^3 f(x)$	$\Delta^4 f(x)$
0	0.00	0.21	0.09	0.06	0.04
1	0.21	0.30	0.15	0.10	−0.03
2	0.51	0.45	0.25	0.07	0.01
3	0.96	0.70	0.32	0.08	−0.02
4	1.66	1.02	0.40	0.06	0.02
5	2.68	1.42	0.46	0.08	
6	4.10	1.88	0.54		
7	5.98	2.42			
8	8.40				

constructed from the data of Table 4.1. Each column (after the second) contains the differences between two of the entries in the column to its left: the entry on the same row and the entry on the row below. Thus, it can be seen that the third column contains entries which are the first forward differences of the respective values of $f(x)$, which are contained in the second column:

$$\Delta f(3) = f(4) - f(3)$$
$$= 1.66 - 0.96$$
$$= 0.70$$

The fourth column contains the second forward differences of $f(x)$ or, which is the same, the forward differences of the entries in the third column, i.e. $\Delta\{\Delta f(x)\}$, or $\Delta^2 f(x)$. Thus, by (4.10),

$$\Delta^2 f(3) = f(5) - 2f(4) + f(3)$$
$$= 2.68 - 2(1.66) + 0.96$$
$$= 0.32$$

Alternatively,

$$\Delta^2 f(3) = \Delta f(4) - \Delta f(3)$$
$$= 1.02 - 0.70$$
$$= 0.32$$

as before.

We will see below that second and higher differences are needed in formulae for interpolation, differentiation and integration. The expressions for these higher differences become increasingly complex. When the calculations are made on a computer there is no disadvantage to using a complex expression; but when the calculations are made by hand, or using only a calculator, a difference table can make the work easier and less subject to error.

A *backward difference table* can be constructed as shown in Table 4.4. As its name implies, it contains the successive backward differences of $f(x)$. The

Table 4.4 A backward difference table based on Table 4.1.

x	$f(x)$	$\nabla f(x)$	$\nabla^2 f(x)$	$\nabla^3 f(x)$	$\nabla^4 f(x)$
0	0.00				
1	0.21	0.21			
2	0.51	0.30	0.09		
3	0.96	0.45	0.15	0.06	
4	1.66	0.70	0.25	0.10	0.04
5	2.68	1.02	0.32	0.07	−0.03
6	4.10	1.42	0.40	0.08	0.01
7	5.98	1.88	0.46	0.06	−0.02
8	8.40	2.42	0.54	0.08	0.02

INTERPOLATION

interesting feature of this table is that the entries are the same as those in Table 4.3, but they appear in different places, i.e. they have different meanings. Thus 0.25 (in the fourth column of each table) is both $\Delta^2 f(2)$ and $\nabla^2 f(4)$.

This equivalence between forward and backward differences can be demonstrated easily using the operator notation. Let f_0 denote $f(x_0)$, where $x = x_0$ is some base value. Then

$$\begin{aligned}
\Delta^{2n} f_{-n} &= \Delta^{2n} E^{-n} f_0 \quad \text{by (4.11)} \\
&= (E - 1)^{2n} E^{-n} f_0 \quad \text{by (4.12)} \\
&= (E - 1)^{2n} E^{-n} E^{-n} f_n \\
&= (E - 1)^{2n} E^{-2n} f_n \\
&= (1 - E^{-1})^{2n} f_n \\
&= \nabla^{2n} f_n \quad \text{by (4.13)}
\end{aligned}$$

The situation with a *central difference table* is more complex since, as we have seen, central differences involve values which do not appear in the table. For example

$$\delta f(3) = f(3.5) - f(2.5)$$

and these two values are not known. However,

$$\begin{aligned}
\delta^2 f(3) &= \delta f(3.5) - \delta f(2.5) \\
&= f(4) - 2f(3) + f(2) \\
&= 0.25
\end{aligned}$$

Two points are to be noted. First, in particular,

$$\delta^2 f(3) = \Delta^2 f(2) = \nabla^2 f(4)$$

and, in general,

$$\delta^{2n} f_0 = \Delta^{2n} f_{-n} = \nabla^{2n} f_n$$

showing that there are, indeed, connections between the entries in central, forward and backward difference tables.

Secondly, *even order* central differences can be expressed in terms of tabular values – values which appear in the original table – but *odd* central differences cannot be so expressed. Nevertheless, as we shall see, central differences are important and useful, and this apparent deficiency can be overcome.

4.4 Interpolation

Interpolation is the process of determining the 'best' estimate of the value of a function $f(x)$ when the value of x is not one of the tabular values. We have some choice in how we decide on the 'best' value. As indicated above, we will approximate the function by an analytical expression and use that to find the desired value. The approximating function can be of any suitable type; moreover, we can choose to fit *all* the data, or only *some* of them.

We often make use of *linear interpolation* because it is easy to do mentally. From Table 4.1 we would say, almost by intuition, that $f(1.5)$ is 0.36. We have implicitly assumed that $f(x)$ varies linearly between $x = 1$ and $x = 2$ (this is the choice of approximating function), and we have neglected the information at all other values except $x = 1$ and $x = 2$ (this is the range of the data over which we have made the approximation).

Other choices are possible. We could force a quadratic to pass through three adjacent points (or, indeed, any three points). We could use the method of least squares to find a function to represent *all* of the data. We could look for more elaborate functions, such as exponential or trigonometric functions, which fit some or all of the data. The list is almost endless.

The usual choice is a *polynomial* interpolation formula (linear, quadratic, cubic, etc.) which is satisfied exactly by several (two, three, four, etc.) adjacent function values in the neighbourhood of the point of interest. The great advantages of a polynomial are that we can find its coefficients quite easily using finite difference operators, and we can easily manipulate the result to achieve not only interpolation but also differentiation and integration.

4.5 Newton's forward formula

Suppose that values of $f(x)$ are given at the equally spaced points $x_0, x_0 \pm h, x_0 \pm 2h$, etc., and that the value of $f(x)$ is required at the point $x_p = x_0 + ph$, where p is any number. Usually p would be positive (i.e. we would choose x_0 to be less than x_p), it will not be very large – probably of the order of unity (i.e. we would not choose x_0 to be *very much* less than x_p) – and it will not be an integer (otherwise x_p would be one of the tabular values and interpolation would not be required). Then

$$\begin{aligned} f(x_p) &= f(x_0 + ph) \\ &= E^p f(x_0) \quad \text{by (4.11)} \\ &= (1 + \Delta)^p f(x_0) \quad \text{by (4.12)} \\ &= \left\{ 1 + p\Delta + \binom{p}{2}\Delta^2 + \cdots \right\} f(x_0) \end{aligned} \quad (4.21)$$

by the binomial theorem, where

$$\binom{p}{i} \equiv \frac{p(p-1)(p-2)\cdots(p-i+1)}{i!}$$

NEWTON'S FORWARD FORMULA

Equation (4.21) is known as *Newton's forward interpolation formula*. Its right-hand side is an infinite series which, to be evaluated, must be truncated after a certain number of terms: the choice is ours, up to the limit imposed by the number of data values available in the table we are using. The forward differences in it must then be expressed in terms of the tabulated values of $f(x)$.

For example, to obtain a quadratic which passes through the three points at x_0, $x_0 + h$ and $x_0 + 2h$, the series in (4.21) is truncated after the third term. Using the notation

$$f_p = f(x_0 + ph)$$

we obtain

$$f_p = f_0 + p\Delta f_0 + \frac{p(p-1)}{2}\Delta^2 f_0 + \text{higher order terms which we neglect}$$

$$= f_0 + p(f_1 - f_0) + \frac{p(p-1)}{2}(f_2 - 2f_1 + f_0) \quad (4.22)$$

$$= \tfrac{1}{2}(f_2 - 2f_1 + f_0)p^2 - \tfrac{1}{2}(f_2 - 4f_1 + 3f_0)p + f_0 \quad (4.23)$$

Equation (4.23) clearly shows that f_p is a quadratic in p; in practice, however, it is just as easy to use (4.22). The fact that these functions pass through the points at x_0, $x_0 + h$ and $x_0 + 2h$ can be verified by giving p the values of 0, 1 and 2 in turn, and seeing that f_p becomes f_0, f_1 and f_2, respectively.

Using linear interpolation on the data in Table 4.1 we saw that $f(1.5) = 0.36$. Quadratic interpolation, using (4.22) with $x_0 = 1$, $p = 0.5$ and the forward differences from Table 4.3 gives

$$f(1.5) = 0.21 + (0.5)(0.30) + \frac{(0.5)(-0.5)}{2}(0.15)$$

$$= 0.21 + 0.15 - 0.01875$$

$$= 0.34125$$

This results from forcing a quadratic through the function values at $x = 1, 2$ and 3. Use of the second forward difference alters the value which is calculated for $f(1.5)$ from that found using linear interpolation (which was 0.36) by the amount -0.01875. Cubic interpolation, in which one more term of (4.21) is retained, contributes a further amount

$$\frac{(0.5)(-0.5)(-1.5)}{2 \times 3}(0.10) = 0.00625$$

suggesting that a 'better' value of $f(1.5)$ is 0.3475. This results from finding (implicitly) the cubic polynomial passing through the tabular values at $x = 1, 2, 3$ and 4.

It must be emphasized that this is really only a *better* value if, in fact, $f(x)$ satisfies a cubic polynomial in the neighbourhood of $x = 1.5$, rather than a quadratic. Indeed, unless we have some theoretical knowledge about the data in the table to guide us, we do not even know that a polynomial is the 'best' type of interpolating function to use. It is just that it is convenient to work with, and generally gives satisfactory results. It should be clear that if a different function, say

$$f(x) = a_1 e^{b_1 x} + a_2 e^{b_2 x} + \cdots$$

were to be assumed, then values of $a_1, b_1, \ldots$ could be found such that the values of $f(x)$ in a nominated part of the table would again be satisfied. In this case the interpolated value of $f(1.5)$ could well be different from any of the values just found. However, the effort involved in the calculation of these quantities $a_1, b_1, \ldots$ would be very much greater than that required by polynomial interpolation formulae, and there is no guarantee that the result would be any more accurate.

Table 4.5 A difficult interpolation problem.

x	0	1	2	3
$f(x)$	1.0	1.0	1.0	1.0

The point is further emphasized by considering the data in Table 4.5. It is 'obvious' that $f(x)$ has the value 1.0 for all values of x (at least between zero and three), but in fact $f(x)$ in this table has been calculated from the formula

$$f(x) = 2^{1/2} \sin(2\pi x + \pi/4)$$

and no interpolation formula could possibly be expected to reveal this from the data in the table. The moral is that we are forced to assume that the tabular data are sufficiently closely spaced for the essential nature of $f(x)$ to be revealed by them. We can never say that we have found the 'best' or 'most accurate' value of $f(x)$ without adding the rider 'subject to the assumption that $f(x)$ is a polynomial of such-and-such a degree over the range considered'.

A further doubt is cast on the accuracy of interpolation if we calculate $f(1.5)$ again, with $x_0 = 0$ and $p = 1.5$, using four terms in the series. This is equivalent to finding a cubic which satisfies the function values at $x = 0, 1, 2$ and 3, and yields

$$f(1.5) = 0.00 + (1.5)(0.21) + \frac{(1.5)(0.5)}{2}(0.09) + \frac{(1.5)(0.5)(-0.5)}{2 \times 3}(0.06)$$

$$= 0.3375$$

which is slightly less than the previous value of 0.3475.

We have now found the following estimates for $f(1.5)$:

linear interpolation, $x_0 = 1$ 0.36
quadratic interpolation, $x_0 = 1$ 0.341 25
cubic interpolation, $x_0 = 1$ 0.3475
cubic interpolation, $x_0 = 0$ 0.3375

and we are entitled to ask which of these several computed values is the most accurate. Is cubic interpolation better than quadratic? Is quadratic better than linear? Although we have not tried to use them, would an exponential function or a trigonometric function be better than any polynomial?

First, in this case the original data were only given to two significant figures in this portion of the table, and we are therefore not really justified in retaining more than two, or at the most three figures in the interpolated quantities. We are nevertheless faced with a selection of values from the various interpolation formulae used – 0.360, 0.341, 0.347 (or 0.348, depending on yet another choice: whether to round down or up!) and 0.337 (or 0.338) – and the questions are still valid. The answer is simply that we do not know. It is tempting to speculate that the last of the computed values – 0.3375, or 0.337 to three significant figures – is the best of those we have found, since it was obtained using two data points below, and two above, the point of interpolation. In contrast, our first attempt at cubic interpolation used the value of $f(x)$ at $x = 4$ (among others). This is more remote from $x = 1.5$, and therefore may be less reliable as a predictor of $f(1.5)$. However, all that we are entitled to say is that *it is probable* that 0.337 is the more accurate if cubic interpolation is used, and it may well be that cubic interpolation is better than quadratic or linear interpolation.

The student may by now be wondering whether the whole exercise is worth the effort! The point which it is hoped will be appreciated is this: while interpolation can yield an *estimate* of the value of a function at a non-tabulated point, no useful limit can be placed on the *accuracy* of that estimate.

Returning yet again to the problem of finding $f(1.5)$, it should be noted that, for these data, nothing is gained by including the fourth difference in the formula. The third differences, in Table 4.3, are almost constant, and the fourth differences are scattered randomly around zero. We can show that if data satisfy an nth degree polynomial *exactly*, then the nth differences will be constant, and the $(n + 1)$th and all higher differences will be zero.

The result follows from the fact (see Table 4.2) that

$$\Delta \equiv e^{hD} - 1$$

Therefore

$$\Delta^n f(x) = (e^{hD} - 1)^n f(x)$$
$$= \left(hD + \frac{h^2 D^2}{2!} + \cdots\right)^n f(x)$$
$$= h^n f^{(n)}(x) + \cdots$$

where the superscript n in parentheses denotes differentiation n times. If $f(x)$ is an nth degree polynomial, then $f^{(n)}(x)$ will be a constant for all x; hence so will $\Delta^n f(x)$. Furthermore $f^{(n+1)}(x)$ and $\Delta^{n+1} f(x)$ will be zero. This characteristic of differences can be used to determine the most appropriate degree of an approximating polynomial for a given set of data and, indeed, whether a polynomial is the most appropriate interpolating function to use.

For example, the fact that the third differences in Table 4.3 are roughly constant suggests strongly that $f(x)$ is a cubic. It is true that the fourth differences are not exactly zero, but they are small and are scattered around zero. We are entitled to assume that they arise from random errors in the original data, and are not truly indicative of the nature of $f(x)$.

If a difference table does not, eventually, contain a column of differences which are, approximately, constant, then the data do not satisfy a polynomial (at any rate, up to the order of differences computed). This does not mean that polynomial interpolation is of no use, but clearly we must be cautious about our expectations of the accuracy of the result.

4.6 Newton's backward formula

We have seen that we can – albeit somewhat hesitantly – find a value for $f(1.5)$ from the data of Table 4.1; but what if we need $f(7.5)$? We could use linear interpolation with $x_0 = 7$, but could not use quadratic interpolation without moving up the table to $x_0 = 6$. Cubic interpolation would require $x_0 = 5$.

An alternative, but essentially equivalent, procedure is to construct an interpolation formula based on backward differences. Since (see Table 4.2)

$$E \equiv (1 - \nabla)^{-1}$$

we have that

$$\begin{aligned} f(x_0 + ph) &= E^p f(x_0) \\ &= (1 - \nabla)^{-p} f(x_0) \\ &= \left\{ 1 + p\nabla + \binom{p+1}{2}\nabla^2 + \cdots \right\} f(x_0) \end{aligned} \quad (4.24)$$

This is *Newton's backward interpolation formula*, and is used in the same manner as (4.21), the forward formula. Thus, to estimate $f(7.5)$ we could set $x_0 = 8, p = -0.5$ and use the differences in Table 4.4 to find that

$$\begin{aligned} f(7.5) &= 8.40 + (-0.5)(2.42) + \frac{(0.5)(-0.5)}{2}(0.54) \\ &\quad + \frac{(1.5)(0.5)(-0.5)}{2 \times 3}(0.08) \\ &= 8.40 - 1.21 - 0.0675 - 0.005 \\ &= 7.1175 \\ &\approx 7.12 \end{aligned}$$

4.7 Stirling's central difference formula

Newton's forward and backward formulae are simple to derive and use, but suffer from the disadvantage that they are 'one-sided' – they use function values which are all above, or all below, the base value. For example, the use of a four-term forward formula to find $f_{1/2}$ involves f_0, f_1, f_2 and f_3. Intuitively, one might expect that it would be better to find a formula involving f_{-1} rather than f_3, since x_{-1} is closer to $x_{1/2}$ and therefore f_{-1} could be expected to be of more relevance to $f_{1/2}$ than f_3 is.

One way of achieving this was suggested in Section 4.5: use f_{-1} as the base value with $p = 1.5$. Another method is to use a *central difference interpolation formula*. Since (see Table 4.2)

$$E \equiv 1 + \delta^2/2 + \delta(1 + \delta^2)^{1/2}$$
$$\equiv 1 + \delta^2/2 + \delta\mu$$

we may express E^p as a power series in δ^2 and $(\mu\delta)$. In place of μ we then use $(1 + \delta^2)^{1/2}$ and expand again. The details are somewhat tedious, and will be omitted (nevertheless, students are encouraged to work through it). The result is

$$f_p = \left(1 + \frac{p^2\delta^2}{2!} + \frac{p^2(p^2-1)\delta^4}{4!} + \cdots\right)f_0$$
$$+ \left(\frac{p\delta}{2!} + \frac{p(p^2-1)\delta^3}{2 \times 3!} + \cdots\right)(f_{1/2} + f_{-1/2})$$

This is known as *Stirling's formula*.

Retaining only the first difference, we obtain

$$f_p = f_0 + (p/2)(\delta f_{1/2} + \delta f_{-1/2})$$
$$= f_0 + (p/2)(f_1 - f_{-1})$$

Applying this to the data of Table 4.1 to find $f(1.5)$ using $x_0 = 1$ yields

$$f(1.5) = 0.21 + (1/4)(0.51 - 0.00) = 0.3375 \qquad (4.25)$$

The contribution of the second difference is

$$(p^2/2)\delta^2 f_0 = (p^2/2)(f_1 - 2f_0 + f_{-1})$$
$$= (1/8)(0.51 - 0.42 + 0.00)$$
$$= 0.011\,25$$

so that now

$$f(1.5) = 0.348\,75 \qquad (4.26)$$

The values in (4.25) and (4.26) can be compared with the values found by the use of Newton's forward formula. Students are recommended to test these various formulae on data for which they can compute the interpolated values exactly; for example, by using trigonometric tables or by generating data from functions of their own choice. Having done that, they are urged *not* to conclude that one or other of the formulae is the 'best', but simply to develop a feel for the probable accuracy of the formulae for data satisfying a variety of functions.

4.8 Numerical differentiation

We turn now to the problem of finding, from tabulated data, the *derivative* of the dependent variable with respect to the independent variable without knowing the functional relationship between the two variables. As in the problem of interpolation, we assume that the data can be represented by a polynomial in the region of interest. We could then construct the polynomial, using one of the interpolation formulae of the previous sections, and differentiate it (analytically) to find the desired derivative. Alternatively, and more directly, we can express $Df(x)$ in terms of differences. This is the approach we shall adopt.

From Table 4.2 we have that

$$hD \equiv \ln(1 + \Delta)$$
$$\equiv \Delta - \Delta^2/2 + \Delta^3/3 - \Delta^4/4 + \cdots$$

Therefore

$$Df(x)_{x=x_n} = f'_n$$
$$= (1/h)(\Delta - \Delta^2/2 + \Delta^3/3 - \Delta^4/4 + \cdots)f_n \qquad (4.27)$$

Approximations involving any desired number of data points are obtained by truncation of (4.27) after the appropriate term. For example, retaining only the first term leads to the two-point forward formula

$$f'_n = (1/h)(f_{n+1} - f_n) \qquad (4.28)$$

The three-point forward formula is

$$f'_n = (1/h)(\Delta - \Delta^2/2)f_n$$
$$= (1/h)\{(f_{n+1} - f_n) - \tfrac{1}{2}(f_{n+2} - 2f_{n+1} + f_n)\}$$
$$= (1/2h)(-f_{n+2} + 4f_{n+1} - 3f_n) \qquad (4.29)$$

and the four-point forward formula is

$$f'_n = (1/6h)(2f_{n+3} - 9f_{n+2} + 18f_{n+1} - 11f_n) \qquad (4.30)$$

Using backward differences, we find that since

NUMERICAL DIFFERENTIATION

$$hD \equiv -\ln(1 - \nabla)$$
$$\equiv \nabla + \nabla^2/2 + \nabla^3/3 + \nabla^4/4 + \cdots$$

the derivative is given by

$$Df_n = f'_n = (1/h)\{\nabla + \nabla^2/2 + \nabla^3/3 + \nabla^4/4 + \cdots\}f_n \quad (4.31)$$

Retaining, respectively, one, two and three terms of (4.31), we obtain

$$f'_n = (1/h)(f_n - f_{n-1}) \quad (4.32)$$
$$f'_n = (1/2h)(3f_n - 4f_{n-1} + f_{n-2}) \quad (4.33)$$

and

$$f'_n = (1/6h)(11f_n - 18f_{n-1} + 9f_{n-2} - 2f_{n-3}) \quad (4.34)$$

Central difference formulae are a little more difficult to obtain. We note that

$$hD \equiv 2\sinh^{-1}(\delta/2)$$

and it can therefore be shown, with some effort (which students are encouraged to make), that

$$hD \equiv \delta - \frac{1}{2^2 \times 3!}\delta^3 + \frac{3^2}{2^4 \times 5!}\delta^5 - \cdots \quad (4.35)$$

However, (4.35) is not convenient to use as it stands, since it involves odd powers of the central difference operator, and hence half-interval values which do not exist. We overcome this problem by use of the identity

$$1 \equiv \mu(1 + \delta^2/4)^{-1/2}$$

which, when applied to (4.35), yields

$$hD \equiv \mu(1 + \delta^2/4)^{-1/2}\left(\delta - \frac{1}{2^2 \times 3!}\delta^3 + \frac{3^2}{2^4 \times 5!}\delta^5 - \cdots\right)$$
$$\equiv \mu\left(1 - \frac{\delta^2}{8} + \frac{3\delta^4}{128} - \cdots\right)\left(\delta - \frac{\delta^3}{24} + \frac{3\delta^5}{640} - \cdots\right)$$
$$\equiv \mu\left(\delta - \frac{\delta^3}{6} + \frac{\delta^5}{30} - \frac{\delta^7}{140} + \cdots\right)$$

Therefore

$$Df_n = f'_n = \left(\frac{\mu}{h}\right)\left(\delta - \frac{\delta^3}{6} + \frac{\delta^5}{30} - \frac{\delta^7}{140} + \cdots\right)f_n \quad (4.36)$$

Retaining only the first term yields

$$f'_n = (\mu/h)\delta f_n$$
$$= (\mu/h)(f_{n+1/2} - f_{n-1/2})$$
$$= (1/2h)(f_{n+1} - f_{n-1}) \quad (4.37)$$

which is the simplest, two-point central difference approximation. Similarly, the retention of two terms of (4.36) leads to

$$f'_n = (1/12h)(-f_{n+2} + 8f_{n+1} - 8f_{n-1} + f_{n-2}) \tag{4.38}$$

Approximations to derivatives higher than the first can readily be found by raising the relevant series to the appropriate power. Thus it will be found that

$$h^2 D^2 \equiv \Delta^2 - \Delta^3 + (11/12)\Delta^4 - (5/6)\Delta^5 + \cdots \tag{4.39}$$

$$\equiv \nabla^2 + \nabla^3 + (11/12)\nabla^4 + (5/6)\nabla^5 + \cdots \tag{4.40}$$

$$\equiv \delta^2 - (1/12)\delta^4 + (1/90)\delta^6 - \cdots \tag{4.41}$$

Note that (4.41) is obtained directly from (4.35), and use of μ is not necessary.

The first terms of (4.39)–(4.41) yield, respectively,

$$f''_n = (1/h^2)(f_{n+2} - 2f_{n+1} + f_n) \tag{4.42}$$

$$= (1/h^2)(f_n - 2f_{n-1} + f_{n-2}) \tag{4.43}$$

$$= (1/h^2)(f_{n+1} - 2f_n + f_{n-1}) \tag{4.44}$$

These three formulae will give, in general, different values for f''_n. Moreover, expression (4.44) is, simultaneously, a *forward* difference formula for f''_{n-1}, a *central* difference formula for f''_n and a *backward* difference formula for f''_{n+1}. These results follow from the fact that the use of a three-point formula implies fitting the data with a parabola – we must expect to get slightly different values from (4.42)–(4.44) unless the data are, exactly, parabolic, and (4.44) gives the *same* value for the second derivative at three *different* neighbouring points because there is only one parabola which passes through three given points and it has the same second derivative at each point.

4.9 Truncation errors

We have seen that we can derive a number of different expressions for the same quantity. For example, (4.28)–(4.34) are all approximations to f'_n, as are (4.37) and (4.38). When evaluated they will, in general, lead to slightly differing values, and they cannot therefore all be correct. Indeed, as we have seen, we cannot be sure that any of them will be correct!

In the first place, they assume that the data can be locally represented by a polynomial of the first, second or higher degree. Since we do not know how true this is, we cannot know how correct the resultant derivatives are. As with interpolation formulae, we must assume, or hope, that the data are sufficiently closely spaced to reveal the true nature of the function, and use a difference table to determine the necessary degree of the polynomial.

TRUNCATION ERRORS

In any case, any expression we use to compute a derivative is obtained by truncating an infinite series. As a result, a *truncation error* is introduced which can be estimated as a function of the first neglected term of the series. Thus (4.28) was obtained by truncating the right-hand side of (4.27) after the first term. The first neglected term is

$$-\Delta^2 f_n/2h \quad (4.45)$$

Since

$$\Delta \equiv e^{hD} - 1$$
$$\equiv hD + h^2D^2/2! + \cdots$$

we may say, to the first order of accuracy,

$$\Delta \equiv hD$$

and hence the truncation error (4.45) becomes

$$-hD^2 f_n/2 \quad (4.46)$$

This expression indicates that the second derivative at $x = x_n$ is involved. A more rigorous analysis shows that the truncation error is

$$-hD^2 f(\xi)/2 \quad (4.47)$$

where ξ is a value of x within the range $x_n \leq \xi \leq x_{n+1}$.

Since we do not know the exact location of ξ, nor can we compute the second derivative in (4.47) – except by numerical means – it would seem as though the information is of little value. However, (4.47) does tell us that the truncation error in (4.28) is proportional to h, so if we could reduce h, then we would reduce the error in the same proportion. We may write

$$f'_n = (1/h)(f_{n+1} - f_n) - (h/2)D^2 f(\xi) \quad (4.48)$$

or, in more general terms,

$$f'_n = (1/h)(f_{n+1} - f_n) + O(h)$$

Similarly, the truncation error of (4.29) is given by

$$f'_n = (1/2h)(-f_{n+2} + 4f_{n+1} - 3f_n) + (h^2/3)D^3 f(\xi) \quad (4.49)$$
$$= (1/2h)(-f_{n+2} + 4f_{n+1} - 3f_n) + O(h^2)$$

It is proportional to h^2, the constant of proportionality being one-third of the third derivative of the function somewhere near the point of interest. That constant cannot be computed, but it is important to know that the truncation error in (4.29) can, approximately, be reduced by a factor of four if h is reduced by a factor of two. [We cannot expect the error reduction to be exactly as predicted by (4.48) or (4.49) because the truncation errors themselves contain errors. However, for sufficiently small values of h it will be found that these error estimates give a good description of the behaviour of the error.]

The various expressions for the truncation error cannot be used to calculate this error and correct the estimate of the derivative, because we cannot obtain a value for the derivative component of the error [for example, $D^3f(\xi)$ in (4.49)] with sufficient accuracy. However, use can be made of the fact that the error is proportional to a known power of the step size h. This is the procedure known as Richardson's extrapolation, and is described in Section 4.16.

4.10 Summary of differentiation formulae

Tables 4.6–4.8* summarize the three-, four- and five-point formulae for the first few derivatives of a function at tabulated points. They extend the results derived above to include formulae which are really not forward, backward or central. For example,

$$\begin{aligned} Df_n &= DEf_{n-1} \\ &= (1/h)\{\ln(1+\Delta)\}(1+\Delta)f_{n-1} \\ &= (1/h)(\Delta - \Delta^2/2 + \Delta^3/3 - \Delta^4/4 + \cdots)(1+\Delta)f_{n-1} \\ &= (1/h)(\Delta + \Delta^2/2 - \Delta^3/6 + \Delta^4/12 - \cdots)f_{n-1} \end{aligned}$$

Retaining three terms we obtain

$$\begin{aligned} Df_n &= (1/h)\{(f_n - f_{n-1}) + \tfrac{1}{2}(f_{n+1} - 2f_n + f_{n-1}) \\ &\quad - (1/6)(f_{n+2} - 3f_{n+1} + 3f_n - f_{n-1})\} \\ &= (1/6h)(-2f_{n-1} - 3f_n + 6f_{n+1} - f_{n+2}) \end{aligned} \qquad (4.50)$$

This gives the derivative at x_n in terms of function values at x_{n-1}, x_n, x_{n+1} and x_{n+2}.

The tables give the coefficients in formulae like (4.48), using the notation

$$D^k f_j = \left(\frac{d^k f}{dx^k}\right)_{x=x_j} = \frac{k!}{h^k n!} \sum_{i=0}^{n} A_{ijkn} f_i + \frac{k!}{h^k} E_{jkn}$$

where k is the order of the derivative;

n is the number of mesh *intervals* used (i.e. $n + 1$ is the number of mesh points);

j is the mesh point at which the derivative is to be evaluated: $0 \leq j \leq n$;

A_{ijkn} is the coefficient of the ith function value;

* Adapted from Bickley, W. G. 1940. Formulae for numerical differentiation. *Mathematical Gazette* **25**, 19.

E_{jkn} is the truncation error of the formula, to be evaluated at some (unknown) point ξ near x_j.

Thus, A_{ijkn} is the coefficient of the ith function value out of the total of $(n + 1)$ which are used to evaluate the kth derivative of $f(x)$ at the point x_j. For example, (4.48) is given by the second line of Table 4.7. (In the tables the subscripts j, k and n are dropped, A_i and E being sufficient, since j, k and n are given explicitly in the column headings.) The error term E is normally a multiple of $h^{n+1}D^{n+1}f(\xi)$; occasionally, for some central difference formulae, E is of a higher order.

Table 4.6 Three-point formulae ($n = 2$).

k	j	A_0	A_1	A_2	E
	0	-3	4	-1	$+\frac{1}{3}h^3 f^{(iii)}$
1	1	-1	0	1	$-\frac{1}{6}h^3 f^{(iii)}$
	2	1	-4	3	$+\frac{1}{3}h^3 f^{(iii)}$
	0	1	-2	1	$-\frac{1}{2}h^3 f^{(iii)}$
2	1	1	-2	1	$-\frac{1}{24}h^4 f^{(iv)}$
	2	1	-2	1	$+\frac{1}{2}h^3 f^{(iii)}$

Table 4.7 Four-point formulae ($n = 3$).

k	j	A_0	A_1	A_2	A_3	E
	0	-11	18	-9	2	$-\frac{1}{4}h^4 f^{(iv)}$
1	1	-2	-3	6	-1	$+\frac{1}{12}h^4 f^{(iv)}$
	2	1	-6	3	2	$-\frac{1}{12}h^4 f^{(iv)}$
	3	-2	9	-18	11	$+\frac{1}{4}h^4 f^{(iv)}$
	0	6	-15	12	-3	$+\frac{11}{24}h^4 f^{(iv)}$
2	1	3	-6	3	0	$-\frac{1}{24}h^4 f^{(iv)}$
	2	0	3	-6	3	$-\frac{1}{24}h^4 f^{(iv)}$
	3	-3	12	-15	6	$+\frac{11}{24}h^4 f^{(iv)}$
	0	-1	3	-3	1	$-\frac{1}{4}h^4 f^{(iv)}$
3	1	-1	3	-3	1	$-\frac{1}{12}h^4 f^{(iv)}$
	2	-1	3	-3	1	$+\frac{1}{12}h^4 f^{(iv)}$
	3	-1	3	-3	1	$+\frac{1}{4}h^4 f^{(iv)}$

Table 4.8 Five-point formulae ($n = 4$).

k	j	A_0	A_1	A_2	A_3	A_4	E
1	0	−50	96	−72	32	−6	$+\frac{1}{5}h^5 f^{(v)}$
	1	−6	−20	36	−12	2	$-\frac{1}{20}h^5 f^{(v)}$
	2	2	−16	0	16	−2	$+\frac{1}{30}h^5 f^{(v)}$
	3	−2	12	−36	20	6	$-\frac{1}{20}h^5 f^{(v)}$
	4	6	−32	72	−96	50	$+\frac{1}{5}h^5 f^{(v)}$
2	0	35	−104	114	−56	11	$-\frac{5}{12}h^5 f^{(v)}$
	1	11	−20	6	4	−1	$+\frac{1}{24}h^5 f^{(v)}$
	2	−1	16	−30	16	−1	$+\frac{1}{180}h^6 f^{(vi)}$
	3	−1	4	6	−20	11	$-\frac{1}{24}h^5 f^{(v)}$
	4	11	−56	114	−104	35	$+\frac{5}{12}h^5 f^{(v)}$
3	0	−10	36	−48	28	−6	$+\frac{7}{24}h^5 f^{(v)}$
	1	−6	20	−24	12	−2	$+\frac{1}{24}h^5 f^{(v)}$
	2	−2	4	0	−4	2	$-\frac{1}{24}h^5 f^{(v)}$
	3	2	−12	24	−20	6	$+\frac{1}{24}h^5 f^{(v)}$
	4	6	−28	48	−36	10	$+\frac{7}{24}h^5 f^{(v)}$
4	0	1	−4	6	−4	1	$-\frac{1}{12}h^5 f^{(v)}$
	1	1	−4	6	−4	1	$-\frac{1}{24}h^5 f^{(v)}$
	2	1	−4	6	−4	1	$-\frac{1}{144}h^6 f^{(vi)}$
	3	1	−4	6	−4	1	$+\frac{1}{24}h^5 f^{(v)}$
	4	1	−4	6	−4	1	$+\frac{1}{12}h^5 f^{(v)}$

4.11 Differentiation at non-tabular points: maxima and minima

The tables in the previous section enable a formula for differentiation to be readily found when the derivative at a tabulated value of x is required, i.e. when p is an integer.

If the derivative at an interpolated point is needed, i.e. at a point other than a tabulated value of x, then a suitable interpolation formula can be differentiated. Thus, from (4.21),

$$f_p = \left(1 + p\Delta + \frac{p(p-1)}{2!}\Delta^2 + \frac{p(p-1)(p-2)}{3!}\Delta^3 + \cdots\right)f_0$$

Therefore

$$\frac{d}{dx}f_p = \frac{d}{dp}f_p \frac{dp}{dx}$$

$$= \frac{1}{h}\left(\Delta + \frac{2p-1}{2}\Delta^2 + \frac{3p^2 - 6p + 2}{6}\Delta^3 + \cdots\right)f_0 \quad (4.51)$$

The coefficients can now be found for a given value of p, including non-integral values, and hence the derivative of a function can be calculated at non-tabular points.

If p is given integer values, then the formulae summarized in Tables 4.6–4.8 are recovered.

Equation (4.51) could also have been obtained directly, thus:

$$\begin{aligned}
Df_p &= DE^p f_0 \\
&= (1/h) \ln(1 + \Delta) \cdot (1 + \Delta)^p f_0 \\
&= (1/h)\left(\Delta - \frac{\Delta^2}{2} + \frac{\Delta^3}{3} - \cdots\right) \\
&\quad \times \left(1 + p\Delta + \frac{p(p-1)}{2!}\Delta^2 + \frac{p(p-1)(p-2)}{3!}\Delta^3 + \cdots\right) f_0 \\
&= (1/h)\left(\Delta + \frac{2p-1}{2}\Delta^2 + \frac{3p^2 - 6p + 2}{6}\Delta^3 + \cdots\right) f_0
\end{aligned}$$

Equations like (4.51) may be used to locate an estimate of the maximum or minimum value of a function which is given only in tabular form.

4.12 Numerical integration

In previous sections, approximations were obtained to the derivative of a function (whether it was explicitly known or not) by establishing a polynomial which was readily differentiated, leading to an estimate of the derivative in terms of tabulated values of the function. We now seek a similar process for approximating the *integral* of a function. The need for a numerical process arises on two counts. First, the function to be integrated may be such that the integral is complicated to evaluate – or may be even impossible to obtain analytically. Secondly, the function may be described only by a table of values, so that a numerical approximation is the only course available.

We wish to find

$$\int_a^b F(x)\,dx$$

where $F(x)$ is a function which either is known but complicated, or is unknown. In either case, we have a table of values of $F(x)$ and the corresponding values of x which are – we will assume – equally spaced along the x-axis from a to b; the spacing will be denoted by h.

Suppose there are M pairs of values of x and $F(x)$. This function is to be replaced by $f(x)$, a polynomial of degree n which matches $F(x)$ at $(n + 1)$ points. The value of $(n + 1)$ may, at most, be equal to M: we cannot compute a polynomial of degree higher than $(M - 1)$ to pass through M points. The coefficients in the polynomial $f(x)$ will, of course, be found from the values of $F(x)$ contained in the table. Once they have been found – or even once expressions for them in terms of the tabular values have been derived – then the polynomial can be integrated and an estimate of the integral of the original function $F(x)$ thereby obtained.

For example, Table 4.1 contains nine pairs of values of x and y. An eighth-degree polynomial may therefore be constructed to pass through these nine points, from which a value can be found for

$$\int_0^8 y(x)\, dx$$

However, we would probably not feel that the data of Table 4.1 should be represented by an eighth-degree polynomial: it has already been suggested that a cubic would be the most suitable polynomial to use. When the data to be integrated contain more than just a few pairs of values of the independent and dependent variable, then the highest order of polynomial that *can* be constructed will generally be of a higher order than we would *wish* to use.

In such a case, the range of x from a to b (from 0 to 8 in Table 4.1) must be divided into two or more sub-ranges, over which separate integrations are performed. The original function $F(x)$ will be represented over each of these sub-ranges by a function $f(x)$ which will be a polynomial of degree $n < M - 1$. The value of n will depend on the number of data points in the sub-range. The extent of the sub-ranges, and the corresponding degrees of the polynomials, need not all be the same, although we would normally seek to keep them approximately equal.

When $F(x)$ denotes a table of values of experimental data, then the size of an interval – or at least the minimum size which can be used – will have been determined when the data were collected. In other cases, however, when the data are generated for the purpose of performing the integration, the choice is freer: we may anticipate that the smaller the interval is, the greater will be the accuracy – but the greater also will be the amount of computation involved.

Suppose, then, that over some portion of the interval (a, b) we seek

$$I_n f(x_0) = \int_{x_0}^{x_0 + nh} f(x)\, dx$$

where the subscript n in I_n denotes that integration is to be performed over n steps along the x-axis.

We need to be able to express the integral operators I and I_n in terms of the other operators. Let $g(x)$ be the *indefinite* integral of $f(x)$, i.e.

$$\int f(x)\, dx = g(x) + \text{constant}$$

or

$$f(x) = Dg(x)$$

or

NUMERICAL INTEGRATION

$$g(x) = D^{-1}f(x)$$

where D^{-1} denotes the operator which is the inverse of D. Then

$$If(x_0) = \int_{x_0}^{x_0+h} f(x)\,dx = g(x_0 + h) - g(x_0)$$
$$= (E - 1)g(x_0) = (E - 1)D^{-1}f(x_0)$$

Thus,

$$I \equiv (E - 1)D^{-1}$$

It follows, therefore, that

$$\int_{x_0+h}^{x_0+2h} f(x)\,dx = EIf(x_0)$$
$$= E(E - 1)D^{-1}f(x_0)$$

Repeating this process of using the shift operator E to move increasingly further along the x-axis, and summing all of the resulting expressions, we obtain

$$I_n f(x_0) = \int_{x_0}^{x_0+nh} f(x)\,dx$$
$$= (1 + E + E^2 + \cdots + E^{n-1})(E - 1)D^{-1}f(x_0)$$
$$= (E^n - 1)D^{-1}f(x_0)$$
$$= \frac{h\{(1 + \Delta)^n - 1\}}{\ln(1 + \Delta)} f(x_0)$$

Whence, after some manipulation (which students are again encouraged to work through), it can be shown that

$$I_n f(x_0) = nh\left(1 + \frac{n}{2}\Delta + \frac{n(2n - 3)}{12}\Delta^2 + \frac{n(n - 2)^2}{24}\Delta^3 \right.$$
$$\left. + n\frac{(6n^3 - 45n^2 + 110n - 90)}{720}\Delta^4 + \cdots\right)f(x_0) \quad (4.52)$$

Various integration formulae follow from (4.52) by giving n different values – i.e. by choosing various sizes for the range (or sub-range) over which to perform the integration – and by retaining different numbers of terms in the series. Some of the more common formulae are given below. To simplify the notation, we use f_0 to denote $f(x_0)$, etc.

$n = 1$, one term:

$$I_1 f_0 = \int_{x_0}^{x_1} f(x)\,dx \approx hf_0 \quad (4.53)$$

which is clearly not a very satisfactory approximation, and would never be used in practice.

$n = 1$, two terms:

$$I_1 f_0 \approx h\left(1 + \frac{\Delta}{2}\right) f_0 = \frac{h}{2}(f_0 + f_1) \qquad (4.54)$$

This is known as the *trapezoidal rule*, and is exact if $f(x)$ is linear. It is simple to use, but not very accurate [except when $f(x)$ is linear, when it would not be needed] unless h is very small.

$n = 2$, three terms:

$$\begin{aligned} I_2 f_0 = \int_{x_0}^{x_2} f(x)\, dx &\approx 2h\left(1 + \Delta + \frac{\Delta^2}{6}\right) f_0 \\ &= \frac{h}{3}(f_0 + 4f_1 + f_2) \end{aligned} \qquad (4.55)$$

This is called *Simpson's rule*. It is exact if the data (i.e. f_0, f_1 and f_2) lie on a quadratic. As it happens, it is also exact for a cubic, since in this case ($n = 2$) (4.52) shows that the coefficient of Δ^3 vanishes. This formula is also simple, and much more accurate than the trapezoidal rule. It is probably the commonest of the numerical integration formulae.

4.13 Error estimation

Before proceeding further we will consider an easy approach to the estimation of the error involved in the integration formulae.

The trapezoidal rule (4.54), for example, is exact if $f(x)$ is linear, since in this situation the second and higher differences are zero. If this is not the case, however, there will be a truncation error which is of the order of the first neglected term, viz. $(-h/12)\Delta^2 f(x)$. Since $\Delta \approx hD$, the error is therefore about $(-h^3/12) f''(\xi)$. The value of ξ at which $f''(\xi)$ should be evaluated is not known, except that it lies within the range of integration. This expression can thus only be used to determine an upper bound for the error – and then only if we have an exact or approximate expression for $f''(x)$. We do learn, however, that the error is approximately proportional to h^3. Errors in other integration formulae may similarly be estimated by examining the first neglected term in the infinite series.

A knowledge of the order of the truncation error is needed to use Richardson's extrapolation (see Section 4.16).

4.14 Integration using backward differences

I_n may also be obtained in terms of backward differences. Since

$$I_n f(x) = (E^n - 1)D^{-1} f(x)$$
$$= \frac{h\{1 - (1 - \nabla)^{-n}\}}{\ln (1 - \nabla)} f(x)$$

we obtain

$$I_n f(x) = nh\left(1 + \frac{n}{2}\nabla + \frac{n(2n+3)}{12}\nabla^2 + \frac{n(n+2)^2}{24}\nabla^3 + n\frac{(6n^3 + 45n^2 + 110n + 90)}{720}\nabla^4 + \cdots\right) f(x_0) \quad (4.56)$$

As with (4.52), we allow n to assume various values and retain various numbers of terms in the series to obtain particular formulae. For example, with $n = 1$ and using three terms we find

$$I_1 f(x) = \int_{x_0}^{x_1} f(x)\, dx = \frac{h}{12} (23 f_0 - 16 f_{-1} + 5 f_{-2}) \quad (4.57)$$

It can be observed that (4.57) evaluates the integral of $f(x)$ over an interval from x_0 to x_1 without using the value of the function at the upper end ($x = x_1$) of that interval. In effect, a polynomial approximation based on the values of x at the lower end ($x = x_0$) and at the two preceding points ($x = x_{-1}$ and $x = x_{-2}$) has been *extrapolated* to enable the integration to be performed. Such a procedure would not normally be used for integration *per se*; there is no need to extrapolate to $x = x_1$ when a value for f_1 is available. However, the technique forms part of the predictor–corrector method of solution of differential equations described in Chapter 5. In that application, forward values of $f(x)$ are not available – they are being calculated only as the solution progresses – and integration formulae using backward differences are used to predict a future value of $f(x)$.

4.15 Summary of integration formulae

Presented here is a collection of some popular integration formulae, together with their respective truncation errors. The student should verify them by the application of the methods outlined above.

$n = 1$ *(trapezoidal rule)*:

$$\int_{x_0}^{x_1} f(x)\, dx = \frac{h}{2}(f_0 + f_1) - \frac{h^3}{12} f^{(ii)}(\xi)$$

$n = 2$ *(Simpson's first or one-third rule)*:

$$\int_{x_0}^{x_2} f(x)\, dx = \frac{h}{3}(f_0 + 4f_1 + f_2) - \frac{h^5}{90} f^{(iv)}(\xi)$$

$n = 3$ (*Simpson's second or three-eighths rule*):

$$\int_{x_0}^{x_3} f(x)\,dx = \frac{3h}{8}(f_0 + 3f_1 + 3f_2 + f_3) - \frac{3h^5}{80}f^{(iv)}(\xi)$$

$n = 4$:

$$\int_{x_0}^{x_4} f(x)\,dx = \frac{2h}{45}(7f_0 + 32f_1 + 12f_2 + 32f_3 + 7f_4) - \frac{8h^7}{945}f^{(vi)}(\xi)$$

$n = 5$:

$$\int_{x_0}^{x_5} f(x)\,dx = \frac{5h}{288}\{19(f_0 + f_5) + 75(f_1 + f_4) + 50(f_2 + f_3)\}$$
$$- \frac{275h^7}{120\,96}f^{(vi)}(\xi)$$

$n = 6$:

$$\int_{x_0}^{x_6} f(x)\,dx = \frac{h}{140}\{41(f_0 + f_6) + 216(f_1 + f_5) + 27(f_2 + f_4) + 272f_3\}$$
$$- \frac{9h^9}{1400}f^{(viii)}(\xi)$$

$n = 6$ (*Weddle's rule*):

$$\int_{x_0}^{x_6} f(x)\,dx = \frac{3h}{10}(f_0 + 5f_1 + f_2 + 6f_3 + f_4 + 5f_5 + f_6) - \frac{h^7}{140}f^{(vi)}(\xi)$$

The last formula is less accurate than the preceding one, but its coefficients are simpler. This may be a slight advantage if computations are being performed using a hand-held calculator. With a digital computer the advantage is negligible, especially when the decrease in accuracy is considered.

These formulae permit the integral of a set of tabulated data to be computed over several intervals of x – from one to six, depending on which formula is used. However, if the data contain more than seven pairs of values of x and y, then these formulae would appear to be inadequate. Formulae which extend over higher numbers of sets of data could be readily obtained, but this would be equivalent to fitting increasingly higher-order polynomials to the data, which would generally not be appropriate.

Instead, as mentioned in Section 4.12, the range of integration is divided into a number of sub-ranges, and one of the foregoing formulae is used repeatedly over each sub-range in turn.

Consider the data in Table 4.1, and suppose that the value of

$$Z = \int_1^7 y\,dx$$

SUMMARY OF INTEGRATION FORMULAE

is required. From the differences in Table 4.3, which was constructed from Table 4.1, we concluded that the best polynomial representation of $y(x)$ is a cubic, which suggests that Simpson's rule should be used for the integration. But Simpson's rule only makes use of three consecutive values of $y(x)$, while to evaluate Z we must integrate y over seven values, i.e. over six intervals of x.

To obtain the result required, we divide the range of integration into a number of sub-ranges, in accordance with the integration formula we have selected. Simpson's (first) rule would suggest

$$Z = \int_1^7 y\,dx = \int_1^3 y\,dx + \int_3^5 y\,dx + \int_5^7 y\,dx$$
$$= \frac{h}{3}(f_1 + 4f_2 + f_3) + \frac{h}{3}(f_3 + 4f_4 + f_5) + \frac{h}{3}(f_5 + 4f_6 + f_7)$$
$$= \frac{h}{3}(f_1 + 4f_2 + 2f_3 + 4f_4 + 2f_5 + 4f_6 + f_7)$$
$$= \tfrac{1}{3}\{0.21 + 4(0.51) + 2(0.96) + 4(1.66) + 2(2.68) + 4(4.10) + 5.98\}$$
$$= 12.85$$

In other words, Simpson's rule is applied repeatedly over the consecutive sub-ranges. In general, Simpson's repeated rule over n equal intervals is

$$\int_{x_0}^{x_n} y\,dx = \frac{x_n - x_0}{6n}(f_0 + 4f_1 + 2f_2 + \cdots + 2f_{n-2} + 4f_{n-1} + f_n)$$

Similar 'repeated' versions of the other integration formulae may be written.

What happens if $\int_1^8 y\,dx$ is required, instead of $\int_1^7 y\,dx$? Simpson's repeated rule clearly only works over an *even* number of intervals, since it uses the intervals in pairs. In this case, Simpson's second rule can be used in conjunction with Simpson's first rule. For example, the range of integration can be subdivided

$$\int_1^8 y\,dx = \int_1^3 y\,dx + \int_3^5 y\,dx + \int_5^8 y\,dx$$

The first two integrals may be found from Simpson's first rule, and the third from Simpson's second rule, which applies over three intervals. The truncation errors of these two formulae are of the same order (h^5), so it is appropriate to use them together like this. An alternative procedure, sometimes used, is to combine Simpson's first rule with the trapezoidal rule: in this example, Simpson's rule would be used three times and the trapezoidal rule once. Since the latter formula has a truncation error of lower order (h^3), it is likely to be less accurate than Simpson's rule. The practice is not recommended for general use.

However, it is worth repeating that we do not *know* how accurate any of these formulae are, because we (normally) do not know the true functional relationship between x and y. So we cannot be certain that the trapezoidal rule will be worse than Simpson's rule. For example, if the function happens to be linear, then they are both exact. Also, for some data it may be true (although this is most unlikely) that a linear approximation is better than a quadratic one.

In general, it is hard to envisage circumstances in which the combination of the Simpson and trapezoidal rules would be better than the combination of Simpson's first and second rules, and the latter should therefore be used in preference.

4.16 Reducing the truncation error

The error involved in any of these formulae for interpolation, differentiation and integration is of the form $E = Ch^m$, where the quantity C includes the term $f^{(m-1)}(\xi)$. Unfortunately, even though we have this information, we cannot reliably calculate the size of the truncation error and correct the value we have obtained from whichever formula is being used. There are two reasons for this.

First, we seldom have an *analytical* expression for $f(x)$, or for its derivatives, and at best we would have to evaluate the derivative numerically – thereby incurring a truncation error in the truncation error. Secondly, we do not know ξ, the value of x at which the derivative has to be calculated. Although we could give ξ some value lying within the range of x over which the formula is being applied, and although we could estimate $f^{(m-1)}(\xi)$ by one method or another, the final value of the truncation error would contain its own error – of an unknown magnitude. A direct use of the truncation error formula is thus not recommended.

However, having said that we cannot calculate $f^{(m-1)}(\xi)$ with any degree of reliability, what we *can* say is that its value – whatever it is – may not vary very much over the range of values of x being used in the particular interpolation, differentiation or integration formula. For a step size of h and a formula involving n intervals along the x-axis starting from the base value x_0, ξ is constrained to lie between x_0 and $x_0 + nh$. If the step size h is small, and if n is also not large, then ξ cannot vary greatly, and therefore neither can $f(\xi)$ and its derivatives.

We will now assume that $f^{(m-1)}(\xi)$ is *constant* over the range of the particular formula (for interpolation, etc.) being used. This assumption improves in quality as both the range and the step size h decrease. By making it, we can increase the accuracy of a finite difference calculation.

The technique may be applied to all three processes – interpolation, differentiation and integration – but is slightly different in the last case from in the first two.

For all three processes, it is necessary to perform the calculations using two different step sizes. In a given table of values of x and $f(x)$, the step size of the table is prescribed and is the smallest (and therefore the most accurate) that can be used. A second, larger step size can be obtained by utilizing only every other entry in the table. If the function values are being generated by some numerical process (such as the solution of a differential equation, as will be described in later chapters), then two different step sizes can readily be obtained.

4.16.1 Interpolation and differentiation

Suppose we use an interpolation or differentiation formula with a truncation error of $E = Ch^m$. If the calculations are done twice, using step sizes of h_1 and h_2, then under the assumption that C is approximately constant the two errors will be related by

$$E_2/E_1 \approx (h_2/h_1)^m \qquad (4.58)$$

If the true value is denoted by V_t and the estimates obtained using h_1 and h_2 are denoted by V_1 and V_2, respectively, then

$$V_t = V_1 - E_1 = V_2 - E_2$$

whence

$$\begin{aligned} E_2 &= (V_2 - V_1) + E_1 \\ &\approx (V_2 - V_1) + E_2(h_1/h_2)^m \\ &\approx \frac{V_2 - V_1}{1 - (h_1/h_2)^m} \end{aligned}$$

Therefore

$$V_t \approx V_2 - \frac{V_2 - V_1}{1 - (h_1/h_2)^m}$$

Because of the approximation involved – C is not really constant – this will not be exactly the true value. Accordingly, we will denote it by V_e, the *extrapolated* value.

In particular, if $h_2 = h_1/2$, then

$$V_e = V_2 + \frac{V_2 - V_1}{2^m - 1} \qquad (4.59)$$

This process is known as *Richardson's extrapolation*, implying that we are seeking to extrapolate the results from two calculations with small but finite step sizes to the limit of zero step size. It would be exact if the quantity C – that is, the derivative in the truncation error term – was truly constant. In general it will not be, and (4.59) can only be expected to yield a better estimate of V_t, not the true value itself.

4.16.2 Integration

The use of Richardson's extrapolation in conjunction with numerical integration is slightly different from that described in the previous subsection. This is because numerical integration requires the *repeated* application of a formula to cover the specified range of values of x, whereas formulae for interpolation and differentiation are *point* formulae, which are applied just once for a given calculation.

Again, we perform an integration using two different step sizes h_1 and h_2. If the integration formula being used has a truncation error of order h^m, then the respective errors *per integration step* will be related by

$$E_2/E_1 = (h_2/h_1)^m$$

However, since $h_2 < h_1$, the use of h_2 will require (h_1/h_2) times as many integration steps to cover the total range of integration. Therefore, the *total* error over the whole range of integration will first be *reduced* by the use of a smaller value of h, but will secondly be *increased* by the need to use more integration steps, and will be given by

$$E_2/E_1 = (h_2/h_1)^m (h_1/h_2) = (h_2/h_1)^{m-1}$$

If the extrapolated value of the integral is denoted by V_e and the estimates obtained using h_1 and h_2 are denoted by V_1 and V_2, respectively, then (4.59) must be amended in the case of numerical integration to

$$V_e = V_2 - \frac{V_2 - V_1}{1 - (h_1/h_2)^{m-1}}$$

In particular, if $h_2 = h_1/2$, then

$$V_e = V_2 + \frac{V_2 - V_1}{2^{m-1} - 1} \tag{4.60}$$

For the commonest of the integration formulae, Richardson's extrapolation takes the following forms:

trapezoidal rule ($m = 3$): $V_e = V_2 + (V_2 - V_1)/3$

Simpson's first and second rules ($m = 5$): $V_e = V_2 + (V_2 - V_1)/15$

Richardson's extrapolation may be applied to any numerical process in which the truncation error is (approximately) proportional to some power of the step size. We will see, in the following chapters, how it may be used to improve the accuracy of the numerical solution of differential equations. It is necessary only to know the order of the truncation error, i.e. the exponent m when the local truncation error is expressed as Ch^m.

Worked examples

1. For the data in Table 4.9, construct a forward difference table and determine the order of the most appropriate polynomial to use for approximating this data.

Table 4.9 Data for the Worked Examples.

x	0	0.2	0.4	0.6	0.8	1.0	1.2
y	-1.651	0.300	2.178	3.505	3.800	2.582	-0.627

The difference table is shown in Table 4.10. It can be seen that the fourth differences are almost zero and the third differences are approximately randomly scattered around the average value of -0.479. If we wish to use a polynomial interpolation formula, we are unlikely to do better than by using a cubic.

Table 4.10 The forward difference table from the data of Table 4.9.

x	y	Δy	$\Delta^2 y$	$\Delta^3 y$	$\Delta^4 y$
0	-1.651	1.951	-0.073	-0.478	-0.003
0.2	0.300	1.878	-0.551	-0.481	0
0.4	2.178	1.327	-1.032	-0.481	0.003
0.6	3.505	0.295	-1.513	-0.478	
0.8	3.800	-1.218	-1.991		
1.0	2.582	-3.209			
1.2	-0.627				

2. Hence find $y(0.25)$ using (a) Newton's forward formula and (b) Stirling's formula.

(a) We will use differences up to the third forward difference, i.e. we will use four function values. Therefore the best base value to use is $x_0 = 0$. This will put two of the four function values below $x = 0.25$, and two above.

In the table, $h = 0.2$. Therefore, at $x = 0.25$,

$$p = (x - x_0)/h = 1.25$$

Whence, from (4.21),

$$f_p = f_0 + p\Delta f_0 + \frac{p(p-1)}{2}\Delta^2 f_0 + \frac{p(p-1)(p-2)}{6}\Delta^3 f_0$$

$$= -1.651 + (1.25)(1.951) + \frac{(1.25)(0.25)}{2}(-0.073)$$

$$+ \frac{(1.25)(0.25)(-0.75)}{6}(-0.478)$$

$$= 0.7950$$

150   INTERPOLATION, DIFFERENTIATION AND INTEGRATION

(b) Using Stirling's formula, and retaining terms up to the third difference, we find

$$f_p = (1 + p^2\delta^2/2)f_0 + \{p\delta/2 + p(p^2 - 1)\delta^3/12\}(f_{1/2} + f_{-1/2})$$

Now,

$$\delta f_{1/2} = f_1 - f_0 \qquad \delta f_{-1/2} = f_0 - f_{-1}$$

therefore

$$\delta(f_{1/2} + f_{-1/2}) = f_1 - f_{-1}$$

Also,

$$\delta^2 f_0 = f_1 - 2f_0 + f_{-1}$$
$$\delta^3 f_{1/2} = f_2 - 3f_1 + 3f_0 - f_{-1}$$
$$\delta^3 f_{-1/2} = f_1 - 3f_0 + 3f_{-1} - f_{-2}$$

therefore

$$\delta^3(f_{1/2} + f_{-1/2}) = f_2 - 2f_1 + 2f_{-1} - f_{-2}$$

Since the value of f_{-2} is needed, we must take $x_0 = 0.4$. Then $p = -0.75$ and

$$\begin{aligned}f_p &= 2.178 + (-0.75)^2\{3.505 - 2(2.178) + 0.300\}/2 \\ &\quad + (-0.75/2)(3.505 - 0.300) + (-0.75/12)\{(-0.75)^2 - 1\} \\ &\quad \times \{3.800 - 2(3.505) + 2(0.300) + 1.651\} \\ &= 0.7949\end{aligned}$$

Which of these values is the more accurate? It is impossible to know.

3. For the data in Table 4.9, find the location and value of $y_{\max}$.

It can be seen that a maximum value of y occurs somewhere between $x = 0.6$ and $x = 0.8$. We can locate this maximum by finding the value of p for which $Df_p = 0$.

We will use (4.51), and truncate it after the third forward difference, since the data can best be represented by a cubic. Choosing $x_0 = 0.4$ as the base value, so that the anticipated location of the maximum is approximately in the middle of the sub-range being used, we obtain

$$Df_p = 0 = \frac{1}{0.2}\left(\Delta + \frac{2p-1}{2}\Delta^2 + \frac{3p^2 - 6p + 2}{6}\Delta^3\right)f(0.4)$$

Thus

$$0 = 1.327 + \frac{2p-1}{2}(-1.032) + \frac{3p^2 - 6p + 2}{6}(-0.481)$$
$$= -0.2405p^2 - 0.5510p + 1.6827$$

This is a quadratic equation for p, of which the relevant solution is $p = 1.737$. The maximum value of y is therefore located at

$$x = x_0 + ph = 0.4 + 1.737(0.2) = 0.747$$

A cubic interpolation formula may then be used to find the value of this maximum. From (4.21), and again using $x_0 = 0.4$, the maximum is found to be

$$f_p = f_{1.737} = f(0.747)$$
$$= 2.178 + 1.737(1.327) + \frac{1.737(0.737)}{2}(-1.032)$$
$$+ \frac{1.737(0.737)(-0.263)}{6}(-0.481)$$
$$= 3.849$$

4. To illustrate the use of Richardson's extrapolation for differentiation, consider the data in Table 4.11. These are values of $y = e^{-x}$, rounded to six

Table 4.11 Data to test Richardson's extrapolation for numerical differentiation.

x	0.2	0.4	0.6	0.8	1.0
$y = e^{-x}$	0.818 731	0.670 320	0.548 812	0.449 329	0.367 879

decimal places. We are therefore in a position to find the 'correct' answer to any numerical process we apply.

We use the data to calculate dy/dx at $x = 0.6$. With a step size $h_1 = 0.4$, and using a central difference approximation of order h^2, we obtain

$$V_1 \approx \frac{y(1.0) - y(0.2)}{0.8} = -0.563\,543$$

and with a step size $h_2 = 0.2$ we find

$$V_2 \approx \frac{y(0.8) - y(0.4)}{0.4} = -0.552\,478$$

Equation (4.59), with $m = 2$, yields

$$V_e = -0.552\,478 + (-0.552\,478 + 0.563\,543)/3 = -0.548\,790$$

The correct value, to six significant figures, is $-0.548\,812$. The errors in V_1, V_2 and V_e are therefore $-0.014\,731$, $-0.003\,666$ and $0.000\,022$, respectively. The extrapolated value is clearly a great improvement. Normally (i.e. not knowing the true value) we would not be able to calculate these errors, but we *would* know – as these values confirm – that E_2 should be approximately equal to $E_1/4$, and that E_e should be much smaller than either E_1 or E_2.

152   INTERPOLATION, DIFFERENTIATION AND INTEGRATION

5. Apply Richardson's extrapolation to the evaluation of

$$\int_{0.2}^{1.0} e^{-x}\, dx$$

from the data of Table 4.11, using the trapezoidal rule and Simpson's (first) rule.

We know the exact value is 0.450 851 (to six significant figures) and can therefore look at the actual errors in different integration formulae. Again, this is not normally the case. The following calculations can be made.

(a) *Trapezoidal rule*, $h_1 = 0.4$:

$$V_{h_1} = (0.4/2)\{(0.818\,731 + 0.548\,812) + (0.548\,812 + 0.367\,879)\}$$
$$= 0.456\,847$$

Trapezoidal rule, $h_2 = 0.2$ (omitting the details):

$$V_{h_2} = 0.452\,353$$

whence

$$V_e = 0.452\,353 + (-0.004\,494)/3 = 0.450\,855$$

(b) *Simpson's rule*, $h_1 = 0.4$:

$$V_{h_1} = (0.4/3)\{0.818\,731 + 4(0.548\,812) + 0.367\,879\}$$
$$= 0.450\,914$$

Simpson's rule, $h_2 = 0.2$:

$$V_{h_2} = 0.450\,855$$

Whence

$$V_e = 0.450\,855 + (-0.000\,059)/15 = 0.450\,851$$

Since we happen to know the correct answer in this case, we can compute the errors in these various estimates. They are:

Trapezoidal rule, $h_1 = 0.4$	1.3%
Trapezoidal rule, $h_2 = 0.2$	0.33%
Trapezoidal rule, extrapolated	0.000 089%
Simpson's rule, $h_1 = 0.4$	0.014%
Simpson's rule, $h_2 = 0.2$	0.000 89%
Simpson's rule, extrapolated	0%

Note that the errors behave in accordance with (4.58) as h is halved.

Problems

1. Derive equations (4.13)–(4.16) and all the relationships in Table 4.2. Show also that (a) $\nabla = \delta E^{-1/2}$ and (b) $\mu\delta = (\Delta E^{-1} + \Delta)/2$.

2. From the following data:

x	0	10	20	30	40	50	60
y	0	80	200	380	500	550	490

estimate $y(31)$ using (a) third-order forward differences and (b) the fourth-order Stirling formula.

3. (a) Construct a difference table for the following data, and hence select the most appropriate interpolating polynomial:

x	0.3	0.4	0.5	0.6	0.7	0.8
y	0.208 19	0.237 77	0.249 58	0.245 82	0.229 20	0.202 76

(b) Estimate $y(0.425)$ using (i) Newton's forward formula and (ii) Stirling's formula.
(c) Find the location and value of y_{max} using (i) Newton's forward formula and (ii) Stirling's formula.

4. The following data have been obtained:

x	0	0.4	0.8	1.2	1.6	2.0	2.4	2.8
y	12.0	15.3	19.1	22.7	25.2	26.0	24.2	19.1

(a) By constructing a difference table, or otherwise, determine the order of the most appropriate polynomial approximation to this data. (b) Using approximations of this order, estimate the following quantities:

(i) y at $x = 1.1$
(ii) dy/dx at $x = 1.6$
(iii) dy/dx at $x = 1.7$
(iv) $\int_0^{2.8} y \, dx$
(v) the location and value of y_{max}

154 INTERPOLATION, DIFFERENTIATION AND INTEGRATION

5. Use the data of Problem 4 to calculate dy/dx at $x = 0.8$ using (a) $h = 0.8$; (b) $h = 0.4$ and (c) Richardson's extrapolation on these values.

6. (a) Show that the derivatives of a function at a non-tabular point may be expressed in terms of backward differences by

$$f'(x) = \frac{1}{h}\left(\nabla + \frac{2p+1}{2}\nabla^2 + \frac{3p^2+6p+2}{6}\nabla^3 + \cdots\right)f(x_0)$$

and derive the corresponding expression for a second derivative.

(b) If $f(x)$ is given by the following values

x	5.0	10.0	15.0	20.0
$f(x)$	1.6094	2.3026	2.7081	2.9957

find values for $f(x)$, $f'(x)$ and $f''(x)$ at $x = 23.8$.

7. Consider the computation of $Z = \int_a^b f(x)\,dx$, and suppose that Z_1 and Z_2 denote the estimates of Z obtained using Simpson's first and second rules, respectively, with step sizes of $h_1 = (b-a)/2$ and $h_2 = (b-a)/3$, respectively.

(a) Show that (subject to appropriate assumptions concerning the truncation errors) a better estimate of Z is given by

$$Z^* \approx 9Z_2/5 - 4Z_1/5$$

(b) If $f(x) = x^{-2}$, $a = 1$ and $b = 3$, find the values of Z_1, Z_2 and Z^*. What are the percentage errors in the three values?

8. Derive Stirling's interpolation formula.

9. Use Stirling's formula to derive an expression for the first derivative of a function, retaining terms up to (and including) the third differences. Hence find, from the following data, the value of x at which $f(x)$ has its maximum.

x	$f(x)$
0.3	0.208 19
0.4	0.237 77
0.5	0.249 58
0.6	0.245 82
0.7	0.229 20
0.8	0.202 76

10. Write a Fortran function for differentiating a set of data which is contained in an array A(I), I = 1,K. Parameters (or COMMON variables) will be

A	the array to be differentiated
K	the number of elements in the vector A
N	the order of the interpolating polynomial to be used
M	the order of the derivative to be computed
IO	the subscript of the base point (x_0, in the notation of the text)
H	h, the step size in x
P	$p = (x - x_0)/h$, defining the point where the differentiation is to be done

If p is an integer, use the appropriate Bickley formula. If p is not an integer, use equation (4.51) or the corresponding formula for a higher derivative. The function should determine whether p is an integer and make the appropriate choice accordingly.

Write a brief main program to demonstrate the use of your function by calculating values of

$$y = \sin x$$

for x in the range (0, 2) in steps of 0.1 and then calculating (a) dy/dx at $x = 0.3$; (b) dy/dx at $x = 0.31$; (c) d^2y/dx^2 at $x = 0.92$ and (d) d^3y/dx^3 at $x = 1.98$.

11. Derive Equation (4.35).

12. Derive Equation (4.52).

13. Derive formulae for the integration of a function in accordance with the following criteria:

(a) $I_1 f(x)$ is a function of f_{-1}, f_0 and f_1
(b) $I_1 f(x)$ is a function of f_{-1}, f_0, f_1 and f_2
(c) $I_3 f(x)$ is a function of f_{-1}, f_0, f_1 and f_2

where, as defined in Section 4.12, $I_n f(x)$ denotes integration over n intervals along the x-axis.

What are the truncation errors in each case?

INTERPOLATION, DIFFERENTIATION AND INTEGRATION

14. The following data were collected during a laboratory experiment:

x	y
0	356.1
0.5	397.1
1.0	423.8
1.5	445.1
2.0	450.7
2.5	509.9
3.0	572.6
3.5	668.2
4.0	806.6

(a) Examination of the data in a graph or difference table will strongly suggest that the value of y at $x = 2$ was incorrectly read or recorded. Using an appropriate interpolation formula, estimate the correct value of $y(2)$.

(b) Having amended the data in accordance with your solution to part (a), find the value of

$$\int_0^4 y \, dx$$

(i) first, using $h = 1$; (ii) then using $h = 0.5$ and (iii) finally, using Richardson's extrapolation.

Justify your choice of integration formula.

5
Ordinary differential equations

5.1 Introduction

In this chapter some methods for the numerical solution of ordinary differential equations are described. First-order equations, of the general form

$$y' = f(x, y) \tag{5.1}$$

or higher-order equations, which can be written

$$y^{(n)} = f(x, y, y', \ldots, y^{(n-1)}) \tag{5.2}$$

have solutions which may be written

$$y = \phi(x) \tag{5.3}$$

The problem is to determine, from (5.1) or (5.2) and the necessary boundary conditions, the relationship (5.3) between x and y.

Analytical methods, when available, generally enable (5.3) to be found explicitly, so that a value of y is known for all values of x. Numerical methods, on the other hand, lead to values of y corresponding only to some finite set of values of x; the solution is obtained as a table of values, rather than as a continuous function. Moreover, (5.3) – if it can be found – is exact, whereas a numerical solution inevitably involves an error which should be small but may, if it is not controlled, swamp the true solution. We must therefore be concerned with two aspects of numerical solutions of ordinary differential equations: both the method itself, and its accuracy.

A common way of handling a second- or higher-order equation is to replace it by an equivalent system of first-order equations. This can be done by the introduction of an intermediate variable $z = y'$, which allows the equation

$$y'' = f(x, y, y')$$

to be replaced by the system

$$y' = z$$
$$z' = f(x, y, z)$$

With a higher-order equation new variables may also be introduced for the second, third, etc., derivatives.

For this reason we shall concentrate almost entirely on methods for the solution of first-order equations.

Since a relationship $y = \phi(x)$ exists between x and y, (5.1) may be written

$$y' = f\{x, \phi(x)\}$$
$$= g(x) \text{ say} \tag{5.4}$$

We do not know $\phi(x)$ – indeed, that is exactly what we are looking for – and therefore $g(x)$ is also not known. Nevertheless, the solution of (5.1) can be formally written

$$y = \int g(x) \, dx + A \tag{5.5}$$

where A is a constant of integration. In order to determine A, an *initial condition* is needed; for example, $y = y_0$ at $x = x_0$. The solution (5.5) then becomes

$$y = y_0 + \int_{x_0}^{x} g(x) \, dx \tag{5.6}$$

We must now consider how the integral of the unknown function $g(x)$ can be found.

We do this in effect – although not explicitly – by expressing $g(x)$ as a polynomial function which satisfies the differential equation and its boundary condition(s) at a finite number of points along the x-axis. The integration in (5.6) may then be performed. Different solution methods result from the use of different polynomial representations of $g(x)$.

5.2 Euler's method

The simplest method for the numerical integration of a first-order ordinary differential equation is Euler's method. This method assumes that, for a small distance Δx along the x-axis from some initial point x_0, the function $g(x)$ is a constant equal to $g(x_0)$.

Consider the problem

$$\frac{dy}{dx} = f(x, y) = g(x) \tag{5.7}$$

with $y = y_0$ at $x = x_0$. It follows that

EULER'S METHOD

$$\left.\frac{dy}{dx}\right|_{x=x_0} = f(x_0, y_0) = g(x_0)$$

If it is assumed that $g(x) = g(x_0)$ for all values of x between x_0 and $x_1 = x_0 + \Delta x$, then the change in y corresponding to the small change Δx in x is given approximately by

$$\frac{\Delta y}{\Delta x} = f(x_0, y_0) \tag{5.8}$$

If y_1 is used to denote $y_0 + \Delta y$, then (5.8) becomes

$$y_1 = y_0 + \Delta x f(x_0, y_0)$$

It is conventional to use the notation h in place of Δx and thus the first step in the solution of (5.7) is given by

$$y_1 = y_0 + hf(x_0, y_0) \tag{5.9}$$

The initial condition provides the values of x_0 and y_0; y_1 can then be calculated from (5.9). This completes the solution process for one step along the x-axis.

The process may now be repeated, using (x_1, y_1) as starting values, to yield

$$y_2 = y_1 + hf(x_1, y_1)$$

and, in general,

$$y_{n+1} = y_n + hf(x_n, y_n) \tag{5.10}$$

This is known as *Euler's method*. The calculations progress, interval by interval, along the x-axis from the initial point x_0 to the required finishing point.

Equation (5.9) can be interpreted geometrically, as shown in Figure 5.1.

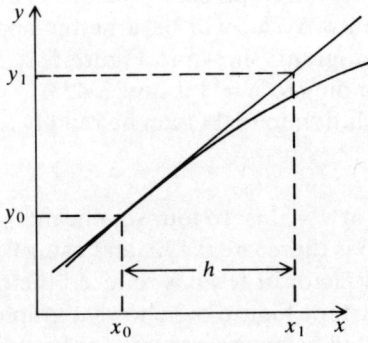

Figure 5.1 Euler's method.

The slope of the curve of y against x is dy/dx, i.e. $f(x, y)$. Thus, at the starting point (x_n, y_n) of each interval the slope is exactly $f(x_n, y_n)$. This value is a moderately good approximation, if h is small, to the average slope over the interval h from x_n to x_{n+1}. Hence, (5.9) yields a moderately good approximation to y_{n+1}. Clearly the approximation

$$\frac{dy}{dx} \approx \frac{\Delta y}{\Delta x}$$

improves as $\Delta x \to 0$. Hence, the accuracy of (5.9) also improves as Δx (i.e. h) is reduced.

As an example, consider the equation

$$\frac{dy}{dx} = x + y \qquad (5.11)$$

with the initial condition $y = 1$ at $x = 0$. Choosing $h = 0.05$, the calculations proceed as follows:

$$\begin{aligned}
y_1 &= y_0 + hf(x_0, y_0) \\
&= y_0 + h(x_0 + y_0) \qquad \text{for the particular problem (5.11)} \\
&= 1 + 0.05(0 + 1) \\
&= 1.05 \\
y_2 &= 1.05 + 0.05(0.05 + 1.05) \\
&= 1.105 \\
y_3 &= 1.105 + 0.05(0.10 + 1.105) \\
&= 1.165
\end{aligned}$$

etc., to four significant figures. The student should verify that y_{20} (at $x = 1$) is 3.307. This is not an arduous task with a hand-held calculator, and is very easy with a programmable calculator.

If h is reduced to 0.005, then the amount of effort is increased ten-fold. Nevertheless, one would hope that greater accuracy per integration step would result, because $\Delta y/\Delta x$ will be a better approximation to dy/dx. A simple computer program, shown in Figure 5.2, allows us to find that the value of y corresponding to $x = 1$ is now 3.423.

The analytical solution to (5.11) can be readily found; it is

$$y = 2e^x - x - 1$$

The true value of y at $x = 1$ is, to four significant figures, 3.437. The error in $y(1)$ using $h = 0.05$ is therefore 0.130, and using 0.005 it is 0.014. Reducing the step size h by a factor of ten has reduced the error at $x = 1$ by a factor which is also about ten. Figure 5.3 shows a graph of the error for various values of h, computed using the program of Figure 5.2. It can be seen that, for a considerable range of values of h, the error is proportional to h.

EULER'S METHOD

The reason that the error is proportional to h can be found by an analysis of *truncation errors*. The Taylor series expansion for the function $y(x)$ about the point x_n is the infinite series

$$y(x) = y_n + \frac{(x - x_n)}{1!} y'_n + \frac{(x - x_n)^2}{2!} y''_n + \cdots$$

which can be written

$$y(x) = \sum_{i=0}^{m} \frac{(x - x_n)^i}{i!} y_n^{(i)} + \frac{(x - x_n)^{m+1}}{(m + 1)!} y^{(m+1)}(\xi) \qquad ((5.12)$$

where $x_n \leq \xi \leq x$ and where the superscript in parentheses denotes differentiation. Euler's method retains only the first two terms in the series, and the final term in (5.12) indicates the magnitude of the *local truncation error*, i.e. the error incurred for each step along the x-axis. If we replace x by $x_n + h$, then (5.12) for Euler's method shows that $y(x_n + h)$, which may be written $y(x_{n+1})$ and also y_{n+1}, is given by

$$y(x_n + h) = y_n + hy'_n + \frac{h^2}{2!} y''(\xi) \qquad (5.13)$$

```
c
c       Define the function f(x,y) and the analytical solution g(x,y)
c          (which, of course, is not normally known):
c
        f(x,y) = x + y
        g(x,y) = 2.*exp(x) - x - 1.
c
c       enter data
c
      1 write (*,100)
    100 format (//' enter initial x, initial y, final x,',
       +          ' step size, and print frequency:')
        read (*,*) x, y, xmax, h, ifreq
        if(h .le. 0.0) stop
        write (*,300)
    300 format(/,'        x           ycalc          true y        error',/)
c
c       solution algorithm starts
c
     30 truey = g(x,y)
        error = truey - y
        write (*,400) x, y, truey, error
    400 format(4f12.4)
        if( x .gt. xmax-h/2. ) go to 1
        icount = 0
     50 y = y + h*f(x,y)
        x = x + h
        icount = icount + 1
        if ( icount .eq. ifreq ) go to 30
        go to 50
c
        end
```

Figure 5.2 The solution of $dy/dx = f(x + y)$.

ORDINARY DIFFERENTIAL EQUATIONS

Typical output from this program appears as follows:

```
enter initial x, initial y, final x, step size, and print frequency:
      .0000          1.0000     1.0000    .0500                  2

     x           ycalc       true y      error

   .0000        1.0000      1.0000      .0000
   .1000        1.1050      1.1103      .0053
   .2000        1.2310      1.2428      .0118
   .3000        1.3802      1.3997      .0195
   .4000        1.5549      1.5836      .0287
   .5000        1.7578      1.7974      .0397
   .6000        1.9917      2.0442      .0525
   .7000        2.2599      2.3275      .0676
   .8000        2.5657      2.6511      .0853
   .9000        2.9132      3.0192      .1060
  1.0000        3.3066      3.4366      .1300

enter initial x, initial y, final x, step size, and print frequency:
      .0000          1.0000     1.0000    .0050                 20

     x           ycalc       true y      error

   .0000        1.0000      1.0000      .0000
   .1000        1.1098      1.1103      .0006
   .2000        1.2416      1.2428      .0012
   .3000        1.3977      1.3997      .0020
   .4000        1.5807      1.5836      .0030
   .5000        1.7933      1.7974      .0041
   .6000        2.0388      2.0442      .0054
   .7000        2.3205      2.3275      .0070
   .8000        2.6422      2.6511      .0089
   .9000        3.0082      3.0192      .0110
  1.0000        3.4230      3.4366      .0135

enter initial x, initial y, final x, step size, and print frequency:
      .0000          .0000      .0000    .0000                  0
```

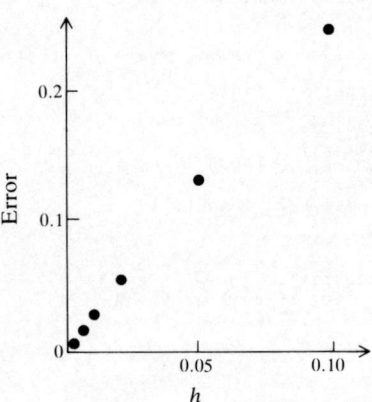

Figure 5.3 The error in Euler's method.

The value of ξ at which the remainder term should be evaluated is not known. At best, therefore, we could only compute an upper bound for the error by estimating the maximum value of $y''(\xi)$ over the range $x_n \leq \xi \leq x_{n+1}$. For present purposes it is sufficient to assume, over a small range of values of x, that y'' is approximately constant, and therefore that the local error – the error per integration step – is proportional to h^2. Since, for a given range of x over which the integration is to be performed, the number of integration steps is proportional to $1/h$, the total or *global* error at the end of the range is proportional to h – the result that was obtained above by numerical experiment.

5.3 Solution using Taylor's series

It can be seen that Euler's method, having an overall error of order h, is not particularly accurate. Despite its simplicity it is seldom used in practice. However, (5.12) immediately suggests how the accuracy could be improved: by retaining more than two terms in the series. Indeed, we may retain as many terms as we wish, it being necessary only to derive expressions for the various higher derivatives involved. This can be achieved with the use of the differential equation being solved. These higher derivatives can be evaluated at the beginning of each interval, just as for the first derivative in Euler's method.

Suppose we wish to solve

$$y' = x^2 - y/x \tag{5.14}$$

with $y = 0.5$ at $x = 1$. The derivatives in (5.12) are obtained from repeated differentiation of (5.14), leading to*

$$y^{(ii)} = (x^2 + 2y/x)/x$$
$$y^{(iii)} = 3(x^2 - 2y/x)/x^2$$
$$y^{(iv)} = -6(x^2 - 4y/x)/x^3$$
$$y^{(v)} = 30(x^2 - 2y/x)/x^4$$

etc. Thus (5.12) becomes, in general,

$$y_{n+1} = y_n + hy_n^{(i)} + \frac{h^2}{2!} y_n^{(ii)} + \frac{h^3}{3!} y_n^{(iii)} + \frac{h^4}{4!} y_n^{(iv)} + \frac{h^5}{5!} y_n^{(v)} + \cdots$$

---

* Although it is conventional to use one or more primes to denote the derivatives of a function, this notation becomes cumbersome when high-order derivatives occur. It is common also to use superscript lower case Roman numerals in parentheses. We shall use both notations without, it is hoped, any confusion.

and, for (5.14) in particular,

$$y_{n+1} = y_n + h\left(x_n^2 - \frac{y_n}{x_n}\right) + \frac{h^2}{2x_n}\left(x_n^2 + \frac{2y_n}{x_n}\right) + \frac{h^3}{2x_n^2}\left(x_n^2 - \frac{2y_n}{x_n}\right)$$
$$- \frac{h^4}{4x_n^3}\left(x_n^2 - \frac{4y_n}{x_n}\right) + \frac{h^5}{4x_n^4}\left(x_n^2 - \frac{2y_n}{x_n}\right) + \cdots \qquad (5.15)$$

Starting at $x_1 = 1$, $y_1 = 0.5$, (5.15) becomes

$$y_2 = y_1 + h\left(x_1^2 - \frac{y_1}{x_1}\right) + \frac{h^2}{2x_1}\left(x_1^2 + \frac{2y_1}{x_1}\right) + \frac{h^3}{2x_1^2}\left(x_1^2 - \frac{2y_1}{x_1}\right)$$
$$- \frac{h^4}{4x_1^3}\left(x_1^2 - \frac{4y_1}{x_1}\right) + \frac{h^5}{4x_1^4}\left(x_1^2 - \frac{2y_1}{x_1}\right) + \cdots$$
$$= 0.5 + 0.5h + h^2 + 0 + h^4/4 + h^5/8 + \cdots$$

We may retain as many terms in this series as we choose to, and give the step size h any value we wish. For example, with $h = 1$ and using the six terms displayed, we obtain

$$y_2 = 0.5 + 0.5 + 1.0 + 0 + 0.25 + 0.125 = 2.375$$

The true value of $y(2.0)$, computed from the analytical solution $y = (x^4 + 1)/4x$, is 2.125. Euler's method, with the same step size, would have retained only the first two terms in (5.15) and led to the value 1.0.

The calculations are now continued from $x_2 = 2$, $y_2 = 2.375$. The student should verify that the next step yields

$$y_3 = 2.375 + 2.8125 + 1.5938 + 0.2031 + 0.0234 + 0.0254 = 7.0332$$

compared with the correct value, which is 6.8333.

With a smaller step size, better accuracy will obviously be obtained. Using $h = 0.5$, for example, the terms involving the fourth and higher derivatives are almost immediately negligible (i.e. affecting only the fourth or later decimal places).

An accurate solution may be obtained from the use of Taylor's series, either by retaining a large number of terms or by using a small step size. In either case, the method clearly offers advantages over Euler's method – which is, of course, the most elementary Taylor series method.

It is clear that a Taylor series solution can lead to an answer with a relatively modest expenditure of effort. However, it will also be clear that this is not always the situation. Consider, for example, this apparently simple problem:

$$y^{(i)} = \frac{1}{x + y} \qquad (5.16)$$

The derivatives necessary for (5.12) must be found from (5.16). The second and third derivatives are

$$y^{(ii)} = \frac{1 + x + y}{(x + y)^3}$$

$$y^{(iii)} = \frac{(x + y)(1 + x + y) - 3(1 + x + y)^2}{(x + y)^5}$$

and it is apparent that higher derivatives will rapidly become very complex – to an extent that their evaluation becomes excessively time-consuming. Since other methods, of equivalent accuracy but smaller computational effort, are available, the Taylor series method is not often used. Nevertheless, it is appropriate for problems in which the derivatives do not become complex, and may also be useful as a starting procedure for predictor–corrector methods which, as discussed below, are not self-starting.

5.4 The modified Euler method

Once

$$y_{n+1} = y_n + hf(x_n, y_n) \tag{5.10}$$

has been used to estimate the value of y_{n+1} at the end of the nth interval, in terms of the slope $f(x_n, y_n)$ at the start of the interval, then the slope $f(x_{n+1}, y_{n+1})$ at the end of that interval can be found. The *average* slope over the interval can next be estimated from

$$\{f(x_n, y_n) + f(x_{n+1}, y_{n+1})\}/2$$

and it seems reasonable to suppose that this quantity will lead to a more accurate value of y_{n+1} through the use of

$$y_{n+1} = y_n + \frac{h}{2}\{f(x_n, y_n) + f(x_{n+1}, y_{n+1})\} \tag{5.17}$$

Equation (5.17) is, generally, a non-linear equation for y_{n+1}, which appears on both sides. It is solved by iteration using (5.10) to provide a first estimate. This method is known as the *modified Euler method*.

For example, using (5.11) again,

$$y' = x + y$$

with $y_0 = 1$ at $x_0 = 0$, and a step size $h = 0.05$, we find first that, as before, (5.10) yields

$$y_1 = y_0 + hf(x_0, y_0)$$
$$= 1 + 0.05(0 + 1)$$
$$= 1.05$$

This is used as the first estimate of y_1, which is now found more accurately by solving (5.17) iteratively. The first iteration yields

$$y_1 = \frac{h}{2}\{f(x_0, y_0) + f(x_1, y_1)\}$$

$$= 1 + \frac{0.05}{2}\{(0 + 1) + (0.05 + 1.05)\} = 1.052$$

Using 1.052 on the right-hand side as the new estimate of y_1, a further iteration yields

$$y_1 = 1 + \frac{0.05}{2}\{(0 + 1) + (0.05 + 1.052)\} = 1.053$$

The iterations have converged to four significant figures, which we may decide are sufficient for our purpose. Starting now from $x_1 = 0.05$, $y_1 = 1.053$, we next estimate y_2 to be 1.108, and then use (5.17) iteratively to improve this estimate and finally find (to four significant figures) the value 1.138.

Continuing until $x = 1$, it is found that the estimate of $y(1)$ is 3.438, compared with the true value of 3.437; the error is thus 0.001. This accuracy is vastly better (see Table 5.1) than that of the simple Euler's method for the same step size ($h = 0.05$), which predicts $y(1) = 3.307$ and an error therefore of -0.130. If the step size is increased to 0.25, then the calculated value of $y(1)$ will be found to be 3.465, which is in error by 0.028. Even with this much larger value of h, the modified Euler method produces a result which is better than that of the simple Euler method with $h = 0.05$. The modification is clearly a great improvement. It should be noted that the error in the modified Euler method has increased by about 25 times (from 0.001 to 0.028) for a five-fold increase in the step size h (from 0.05 to 0.25). It will be shown below that the global truncation error for this method is of order h^2.

Equation (5.17) is the important stage of the modified Euler method; (5.10) is used merely to provide a first estimate of y_{n+1}, a step which is needed because (5.17) is (in general) a non-linear equation which is solved iteratively. In principle, other first estimates may be used: for example the converged value of y_n, although this obviously would not be a very good first estimate. Since the number of iterations needed to reach convergence of (5.17) depends on the quality of the first estimate, it should be as good as

Table 5.1 A comparison of the errors in some methods for the solution of (5.11).

Method	Error in $y(1)$
simple Euler method with $h = 0.05$	-0.130
simple Euler method with $h = 0.005$	-0.014
modified Euler method with $h = 0.25$	0.028
modified Euler method with $h = 0.05$	0.001

possible, and the small amount of extra work involved in the use of (5.10) is worthwhile.

It will be recalled from Section 2.5 that the condition under which the equation

$$y = F(y)$$

converges when solved by simple iteration is

$$|F'(S)| < 1 \tag{2.22}$$

where S is the solution. Applying this to (5.17), which is to be solved for y_{n+1}, and noting that the right-hand side may be written

$$F(y_{n+1}) = \left\{ y_n + \frac{h}{2} f(x_n, y_n) \right\} + \frac{h}{2} f(x_{n+1}, y_{n+1})$$

we see that convergence requires that

$$\left| \frac{h}{2} f'(x_{n+1}, S) \right| < 1 \tag{5.18}$$

where S is the true value of y at x_{n+1} and *the prime denotes differentiation with respect to y*. Equation (5.18) is used to determine the *upper limit* to the step size h which will permit (5.17) to converge.

Normally, of course, S is not known, so implementation of (5.18) as it stands is not possible. In addition, $f'(x, y)$ will vary along the solution curve, so that the maximum allowable value of h will also vary as the solution proceeds, whereas a constant step size h is often preferred. In applying (5.18), therefore, it may be necessary to estimate the solution roughly (using the simple Euler method, for example), and hence estimate $f'(x, y)$ over the entire solution range. The largest (absolute) value is then used in (5.18) to determine the limit on h which will allow the modified Euler method to converge at all values of x in the solution range.

In the case of (5.11) the determination of h_{max} is very simple. Since

$$f(x, y) = x + y$$

therefore

$$f'(x, y) = 1$$

(Remember that here the differentiation is with respect to y, and x is a constant.)

Equation (5.18) then tells us that convergence of (5.17) requires

$$\frac{h}{2} (1) < 1$$

or

$$h < 2$$

Any value of h up to this limit will allow (5.17) to converge.

The largest value of h permissible for convergence is not necessarily the value which would be used. There are two reasons for this. First, we want (5.17) not merely to converge, but to converge in just a few iterations. The rate of convergence is given by $|hf'/2|$, and we would therefore want to keep h small. Secondly, the accuracy of the solution must also be considered, and will increase as h is reduced. There are thus altogether three aspects to the determination of h: whether the solution can be obtained at all; whether the solution can be obtained with an acceptable computing cost; and whether the truncation error is acceptable.

5.5 Predictor–corrector methods

The modified Euler method is an example of a more general class of methods known as predictor–corrector (P–C) methods. Equation (5.10) is a 'predictor', and (5.17) is a 'corrector'. These methods are derived from the integration formulae of Chapter 4 in a fairly simple manner.

The formal integration of

$$\frac{dy}{dx} = f(x, y)$$

over the range x_{n-j} to x_{n+1} (where j is any integer) may be written

$$y_{n+1} = y_{n-j} + \int_{x_{n-j}}^{x_{n+1}} f(x, y)\, dx \tag{5.19}$$

Since y is a function of x – indeed, $y(x)$ is the solution sought – it is proper in (5.19) to refer to the integration of $f(x, y)$ with respect to x. Introducing the notation

$$p = (x - x_n)/h$$

(5.19) becomes

$$y_{n+1} = y_{n-j} + h \int_{-j}^{1} f_p \, dp \tag{5.20}$$

where

$$f_p \equiv f\{p, y(p)\}$$
$$= f\{x_n + hp, y(x_n + hp)\}$$

Now,

$$f_p = E^p f_n$$
$$= (1 - \nabla)^{-p} f_n$$
$$= \{1 + p\nabla + \tfrac{1}{2}p(p + 1)\nabla^2 + \cdots\} f_n$$

Therefore

$$\int_{-j}^{1} f_p \, dp = \left((1+j) + \frac{1-j^2}{2} \nabla + \frac{5 - 3j^2 + 2j^3}{12} \nabla^2 + \cdots \right) f_n$$

and hence

$$y_{n+1} = y_{n-j} + h\left((1+j) + \frac{1-j^2}{2} \nabla + \frac{5 - 3j^2 + 2j^3}{12} \nabla^2 + \cdots \right) f_n \quad (5.21)$$

By choosing a value for j and truncating the series (5.21) after a selected number of terms, a variety of integration formulae can be obtained. These allow y_{n+1} to be computed in terms of y_{n-j}, h, f_n and one or more backward differences of f_n. The formulae can be of any desired order of accuracy, and can cover any number of intervals along the x-axis.

For example, the choice of $j = 0$ and the retention of only one term inside the parentheses of (5.21) leads to

$$y_{n+1} = y_n + hf_n$$

i.e. the Euler method. In general, for $j = 0$

$$y_{n+1} = y_n + h(1 + \tfrac{1}{2}\nabla + \tfrac{5}{12}\nabla^2 + \cdots) f_n \quad (5.22)$$

However many terms in (5.22) are retained, y_{n+1} will always be given in terms of f_n and backward differences of f_n; i.e. in terms of quantities which do not involve y_{n+1}. Thus, if a series of values of x and y up to and including x_n and y_n are available, then y_{n+1} can be computed, i.e. *predicted*, using (5.22).

A *corrector* formula, such as (5.17), involves y_{n+1} on the right-hand side as well as the left-hand side. To obtain such a formula, backward differences based on f_{n+1} are required. Thus, (5.20) must be replaced by

$$y_{n+1} = y_{n-j} + h\mathrm{B}(\nabla) f_{n+1} \quad (5.23)$$

where $\mathrm{B}(\nabla)$ is an unknown function of the backward difference operator which has to be determined. Now,

$$y_{n+1} = \mathrm{E} y_n$$

and

$$y_{n-j} = \mathrm{E}^{-j} y_n$$

Also,

$$f_{n+1} = \mathrm{E} f_n$$
$$= \mathrm{E} \mathrm{D} y_n \quad \text{by (5.1)}$$

Thus, (5.23) can be written

$$\mathrm{E} y_n = \{ \mathrm{E}^{-j} + h\mathrm{B}(\nabla) \mathrm{E} \mathrm{D} \} y_n$$

and since $\mathrm{E} = (1 - \nabla)^{-1}$ and $h\mathrm{D} = -\ln(1 - \nabla)$, it follows that

$$B(\nabla) = \frac{1 - (1 - \nabla)^{j+1}}{-\ln(1 - \nabla)}$$

It can be shown [by expanding $\ln(1 - \nabla)$ as a power series and cross-multiplying] that

$$\frac{1}{-\ln(1 - \nabla)} = \frac{1}{\nabla} - \frac{1}{2} - \frac{\nabla}{12} - \frac{\nabla^2}{24} - \cdots$$

and hence that

$$B(\nabla) = (j + 1)\left(1 - \frac{j+1}{2}\nabla + \frac{2j^2 + j - 1}{12}\nabla^2 - \frac{(j+1)(j-1)^2}{24}\nabla^3 + \cdots\right)$$

Finally, (5.23) becomes

$$y_{n+1} = y_{n-j} + h(j+1)\left(1 - \frac{j+1}{2}\nabla + \frac{2j^2 + j - 1}{12}\nabla^2 - \frac{(j+1)(j-1)^2}{24}\nabla^3 + \cdots\right)f_{n+1} \quad (5.24)$$

As before, a value is chosen for j and the series is truncated. An estimate of y_{n+1} is thus given in terms of y_{n-j}, h, f_{n+1} and backward differences of f_{n+1}. Using (5.24), a corrector formula of any desired order of accuracy, and covering any desired number of steps along the x-axis, can be constructed.

For example, the choice of $j = 0$ and the retention of two terms inside the parentheses of (5.24) lead to

$$y_{n+1} = y_n + h(1 - \nabla/2)f_{n+1}$$
$$= y_n + (h/2)(f_n + f_{n+1})$$

i.e. the modified Euler corrector. In general, for $j = 0$,

$$y_{n+1} = y_n + h(1 - \tfrac{1}{2}\nabla - \tfrac{1}{12}\nabla^2 - \cdots)f_{n+1} \quad (5.25)$$

Equation (5.25) shows that the truncation error of the modified Euler corrector – which is the first neglected term in the series – is $(-h/12)\nabla^2 f_{n+1}$. Using $\nabla \approx hD$, this is equivalent to $(-h^3/12)D^2 f_{n+1}$ and, in particular, the local error of the modified Euler corrector is proportional to h^3. Thus, the global error, after a number of steps which is proportional to h^{-1}, is proportional to h^2. This is consistent with the behaviour found in Section 5.4.

5.6 Milne's method, Adams' method, and Hamming's method

The values of j used in (5.21) and (5.24) need not be zero, nor do they need to be the same in each equation. A well-known P–C method, known as Milne's method, uses a three-term predictor with $j = 3$ and a three-term corrector with $j = 1$. The resulting formulae are

$$y_{n+1} = y_{n-3} + h\left(4 - 4\nabla + \frac{8}{3}\nabla^2\right)f_n$$

$$= y_{n-3} + \frac{4h}{3}(2f_n - f_{n-1} + 2f_{n-2}) \tag{5.26}$$

and

$$y_{n+1} = y_{n-1} + h\left(2 - 2\nabla + \frac{1}{3}\nabla^2\right)f_{n+1}$$

$$= y_{n-1} + \frac{h}{3}(f_{n+1} + 4f_n + f_{n-1}) \tag{5.27}$$

It should be noted that both (5.26) and (5.27) are of a higher order of accuracy than appears at first glance. They each use three terms in the respective series (5.21) and (5.24). However, if the next terms – the terms in ∇^3 – are derived, then it will be found that their coefficients are zero in each case, because of the particular values chosen for j. The truncation errors of each formula are, in fact, proportional to $h\nabla^4 f$, i.e. they are of order h^5.

The predictor equation (5.26) is to be used once to provide an estimate, y^p say, of y_{n+1}. This estimate is then corrected by solving the corrector equation (5.27) iteratively, using y^p as the first 'guess' at the solution. The successive 'corrected' estimates of y_{n+1} can be denoted $y^{c,1}, y^{c,2}, \ldots, y^{c,k}$ and (5.27) can then be written

$$y_{n+1}^{c,k+1} = y_{n-1} + \frac{h}{3}(f_{n+1}^{c,k} + 4f_n + f_{n-1})$$

We must again ensure that this iterative procedure applied to the corrector will converge. The condition analogous to (5.18) for the convergence of (5.27) is

$$\left|\tfrac{1}{3}hf'(x_{n+1}, S)\right| < 1 \tag{5.28}$$

where S is the (unknown) true value of y at $x = x_{n+1}$. Equation (5.28) permits h to be 50% larger than the value allowed by (5.18). In practice, for computational efficiency, we would require not merely that (5.28) be satisfied, but that it be well and truly satisfied: i.e. that $|hf'/3| \ll 1$. This is necessary to ensure that (5.27) converges in three or four iterations. Otherwise, other methods such as the Runge–Kutta methods (see Section 5.9) may be superior.

Another common corrector formula, known as Adams' method, is obtained from (5.24) by retaining four terms and setting $j = 0$. It is

$$y_{n+1} = y_n + \frac{h}{24}(9f_{n+1} + 19f_n - 5f_{n-1} + f_{n-2}) \tag{5.29}$$

Milne's predictor (5.26) may be used to provide a first estimate of y_{n+1} to use in the first evaluation of f_{n+1}. (*Any* predictor may be used, but one which is more accurate will cause fewer iterations of the corrector to be needed.) The truncation error of Adams' method is, like that of (5.27), of order h^5. It appears to require more work than Milne's method does, because there seem to be four evaluations of the function $f(x, y)$ per iteration in the former and only three in the latter. In fact, the only function evaluation required each iteration is that of f_{n+1}: values of f_n, f_{n-1}, etc., will have been obtained in previous time steps and should have been stored for future use.

A more extensive class of P–C methods, of which the general form is

$$y_{n+1} + \beta_0 y_n + \beta_1 y_{n-1} + \cdots + \beta_j y_{n-j}$$
$$= h(\alpha_{-1} f_{n+1} + \alpha_0 f_n + \alpha_1 f_{n-1} + \alpha_2 f_{n-2} + \cdots + \alpha_k f_{n-k}) \quad (5.30)$$

can be derived; the details are omitted here. A useful member of this class is Hamming's method

$$y_{n+1} = \frac{1}{8}(9y_n - y_{n-2}) + \frac{3h}{8}(f_{n+1} + 2f_n - f_{n-1}) \quad (5.31)$$

Again, Milne's predictor (or any other) may be used first. Hamming's method also has a truncation error of order h^5.

5.7 Starting procedure for predictor–corrector methods

Predictor–corrector methods (other than the modified Euler method) possess one obvious drawback: they are not 'self-starting'. The use of (5.26), for example, implies a knowledge of the values of y at x_{n-3}, x_{n-2} and x_{n-1}, as well as at the current point x_n. Invariably, these data will not be provided in the initial conditions of the problem, which (for a first-order differential equation) will give only one set of values of x and y: say (x_0, y_0).

Starting values must be obtained by using a different method. For example, the modified Euler method (which *is* a single-step method) can be used to obtain $(x_1, y_1), (x_2, y_2)$ and (x_3, y_3). However, for a given mesh size it is relatively inaccurate and, as a consequence, the further values calculated using Milne's (or any other higher order) method would be based upon poor quality starting data: they would have what is called an inherited error, and would therefore themselves be in error. To overcome this a smaller step size must be used while finding the starting values.

The modified Euler method has a local truncation error – a truncation error per step – which is proportional to the cube of the step size; i.e. to h_E^3, say. Since the number of steps needed to integrate over a given range of x is proportional to $1/h_E$, such an integration will incur a global error which is of the order of h_E^2. To cover the same range using (for example) Milne's method, which has a local error of order h_M^5, will give rise to a global error which is proportional to h_M^4. Of course, the constants of proportionality – the coefficients in the respective truncation errors – are not the same in the two cases, but leaving that aside, this argument suggests that h_E should be roughly proportional to h_M^2. An examination of the actual coefficients for a given differential equation will indicate more clearly what the most suitable choice for h_E is likely to be.

The coefficient of the truncation error includes a derivative of the function $f(x, y)$ appearing in the original differential equation. To estimate the truncation error therefore requires a knowledge of the solution, which must either be 'guessed' or obtained from a quick and less accurate numerical solution.

For example, consider the equation

$$y' = \cos x - \sin x + y = f(x, y)$$

with $y = 0$ at $x = 0$.

Equation (5.24) shows that the truncation error of the modified Euler method is $h\nabla^2 f_{n+1}/12 = h^3 f^{(iii)}/12$, and that of Milne's method is $h\nabla^4 f_{n+1}/90 = h^5 f^{(v)}/90$, where the differentiation is with respect to x and where the derivatives are to be evaluated at some unknown point within the current step. We find that $f^{(iii)}(x, y)$ in this case happens to be simply y and that $f^{(v)}(x, y)$ is $y - \cos x - \sin x$. With the given initial condition we would find, using the simple Euler method (or perhaps by intuition) that the solution (i.e. the values of y) will be of unit order of magnitude for all values of x. Therefore $f^{(iii)}$ and $f^{(v)}$ will each also be of order of magnitude 1.

The local truncation error of the modified Euler method will thus be about $h^3/12$, and that of Milne's method will be about $h^5/90$, and the respective global errors will be $h^2/12$ and $h^4/90$. We obtain finally that h_E should be about $h_M^2/2.7$ (for this particular equation).

Assuming, for example, that we select $h_M = 0.1$, then h_E should be about 0.0037. Clearly we would wish to round that off to a more convenient value. To be on the safe side, we could choose $h_E = 0.0025$ (i.e. we take a step size smaller than the indicated value, instead of larger).

Milne's method requires that we compute values of y at $x = 0.1, 0.2$ and 0.3, to be able to continue the solution to $x = 0.4$ (and beyond). Using the modified Euler method with $h_E = 0.0025$, values of y from $x = 0.0025$ to $x = 0.3$ can be calculated. The values of y at $x = 0.1, 0.2$ and 0.3 are extracted from the solution table for use in the continuation of the solution. The intermediate values, at $x = 0.0025, 0.005, \ldots, 0.0975, 0.1025, \ldots, 0.1975, 0.2025, \ldots, 0.2975$, can be discarded.

Starting values for a predictor–corrector method can also be obtained using a Taylor series solution. Although, as discussed above, this method may involve the evaluation of very complex functions, and hence be very time-consuming, the cost will not be great if only a few solution steps are needed. The great advantage of using the Taylor series method to find starting values is that the truncation error can be both fairly well-controlled (by taking additional terms) and estimated; hence the P–C method will not start off with an excessive and unknown inherited error. A further advantage is that sufficient terms may be retained in the Taylor series (if the derivatives are not too unwieldy) to give it the same truncation error, or at least the same order of truncation error, as the P–C method, so that the same step size may be used.

5.8 Estimation of error of predictor–corrector methods

An estimate of the error of a P–C step – as distinct from an estimate merely of the *order* of the error – can be obtained from the truncation errors of the predictor and corrector formulae, whenever these are of the same order in terms of h. Consider Milne's method as an example. The predictor and corrector formulae may be written

$$y_{n+1} = y_{n-3} + \frac{4h}{3}(2f_n - f_{n-1} + 2f_{n-2}) + \frac{14}{45} h\nabla^4 f\{\xi_1, y(\xi_1)\}$$
$$= y^p + \frac{14}{45} h\nabla^4 f\{\xi_1, y(\xi_1)\} \tag{5.32}$$

and

$$y_{n+1} = y_{n-1} + \frac{h}{3}(f_{n+1} + 4f_n + f_{n-1}) - \frac{1}{90} h\nabla^4 f\{\xi_2, y(\xi_2)\}$$
$$= y^c - \frac{1}{90} h\nabla^4 f\{\xi_2, y(\xi_2)\} \tag{5.33}$$

where y^p and y^c are the values given by (5.26) and (5.27). The values of the quantities ξ_1 and ξ_2 lie somewhere in the interval from x_n to x_{n+1} and to proceed further we must now assume that they are sufficiently close to each other to allow the approximation

$$\nabla^4 f\{\xi_1, y(\xi_1)\} = \nabla^4 f\{\xi_2, y(\xi_2)\} = \nabla^4 f \quad \text{(say)} \tag{5.34}$$

to be made. Then (5.32) and (5.33) may be combined by subtraction to yield

$$\tfrac{1}{90} h\nabla^4 f = \tfrac{1}{29}(y^c - y^p) \tag{5.35}$$

Equation (5.35) provides an estimate of the error in the converged solution of (5.33) which can be used to determine whether, for example, the step size requires adjustment or whether formulae of higher accuracy should be used. Because of the somewhat doubtful validity of (5.34) – it may be a good approximation in some cases, and a poor one in others – (5.35) should not be used actually to improve on the value of y^c.

Similar estimates can be obtained for Adams' and Hamming's methods, the respective corrector truncation errors of which are

$$-\tfrac{19}{720} h\nabla^4 f\{\xi_2, y(\xi_2)\} \quad \text{and} \quad -\tfrac{1}{40} h\nabla^4 f\{\xi_2, y(\xi_2)\}$$

The derivation of such estimates is left as an exercise for the student.

Milne's method has a weakness – it may be unstable. Small errors – round-off errors – may grow without limit and become large compared with the solution. The equation $y' = \alpha y$ (with $y = 1$ at $x = 0$) is a simple problem which highlights this possibility.

When α is positive, there is no difficulty: the solution itself grows exponentially, and although the error in the numerical solution also grows, it remains within bounds. For example, with $\alpha = 1$ and using a step size $h = 1$, the numerical solution at $x = 10$ by Milne's method is $y = 22\,824$ compared with the true value of $22\,026$. The error is about 817. By comparison, the errors in solutions obtained using Adams' and Hamming's formulae are 2123 and 2188, respectively*.

However, when α is negative the solution decays exponentially to zero – but the error does not. For example, with $\alpha = -1$ and $h = 1$, Milne's method leads to $y(20) = -0.0398$, compared with the analytical value of 2×10^{-9}. Adams' and Hamming's methods both yield numerical solutions of zero. At $x = 50$, Milne's method gives the disastrous result $y = 459.8$; Adams' and Hamming's methods are still giving the correct answer.

With more general and more complex problems, it may not be clear without numerical experiments which method will be the best. In all cases the step size should be kept small, to ensure that the corrector will converge rapidly and the truncation error is acceptable. Also, with Milne's method this *generally* means that the instability will not be a problem, as it can always be delayed by reducing the step size; and reduction in step size is always a good practice to ensure that the numerical solution is independent of step size (to within whatever tolerance is appropriate for the particular problem).

However, adjustment of step size may not be easy with P–C methods, because they are what are known as *multi-step* methods: they require the knowledge of a sequence of values of $f(x, y)$ in order that a new value of y can be computed. For example, (5.32) requires values of f_{n-2}, f_{n-1} and f_n to be known in order to predict f_{n+1} and hence y_{n+1}. Methods which do not require a knowledge of previous history (i.e. f_{n-1}, f_{n-2}, etc.) but only of the current values of x and y (i.e. only of the value of f_n at the beginning of each interval in x) are known as *single-step* methods†.

Suppose that the step size is, for some reason, to be *doubled*‡ and also that a sufficient number of steps has already been computed (and the results saved). Then there is no difficulty about doubling the step size: the required 'previous values' will be available, and it is only necessary to ensure that the correct ones are extracted.

---

* These results were obtained using *exact* values, computed from the analytical solution $y = e^x$, as the starting values for the respective P–C method.

† Although the modified Euler method was shown to be the simplest example of a P–C method, it is, in fact, a single-step method: it uses $j = 0$ in (5.21) and (5.24), and retains no terms in the series beyond those involving f_n and f_{n+1}. Any other P–C method is a multi-step method.

‡ Perhaps because (5.35) indicates that the error may be much smaller than is necessary, and therefore that computer time (and money) is being wasted.

On the other hand, if, to obtain improved accuracy, the step size must be *halved*, then some of the required function values will be missing. For example, if solutions have been obtained at $x = 1.0, 1.2, 1.4$ and 1.6, using Milne's method, and (5.35) suggests that the step size should now be halved, then the next solution at $x = 1.7$ will require function values at $x = 1.3, 1.4, 1.5$ and 1.6. The values at $x = 1.3$ and 1.5 have not been calculated, and will need to be obtained by interpolation. This can be done, but it is important to use an interpolation formula which is at least of the same order of accuracy as the integration formula itself, otherwise errors will be introduced which will propagate through the rest of the solution.

5.9 Runge–Kutta methods

The names of Runge and Kutta are traditionally associated with a class of methods for the numerical integration of ordinary differential equations which differ from P–C methods in two significant characteristics.

(a) They are single-step methods. The integration over one interval is self-contained, so that no special starting procedure is required; this also means that the interval may readily be changed as dictated by the form of the solution.

(b) In most cases an estimate of the error can *not* be made (although the *order* of the error in terms of step size is, as usual, known).

It therefore seems that Runge–Kutta (R–K) methods possess some advantages, and some disadvantages, compared with P–C methods. It is not possible to reach any general conclusion as to which method is 'better'. High-accuracy P–C methods can be generated more readily than the corresponding R–K methods. On the other hand, the ease with which the step size can be changed, combined with the procedure for error estimation available in the Runge–Kutta–Merson (R–K–M) method, to be described below, makes this technique attractive. Both methods have their disciples, and both methods can be associated with certain types of equations to which they are more suited. Often a choice is immaterial, and often the choice is merely a matter of preference or convenience. It is suggested that students should acquire experience with each method and reach their own conclusions.

Runge–Kutta methods make several (commonly four or five) estimates of the change in the value of y corresponding to a specified change in x. The value finally adopted is a *weighted average* of these estimates, with the weighting factors chosen so as to minimize the error.

The formal derivation of the formulae is somewhat lengthy, and will be omitted. We will content ourselves with a description of two examples of these methods, including some discussion which sets out the rationale behind all R–K methods. What we shall *not* do is derive – or even justify our choice of – the values of the weighting factors. They will merely be stated.

Suppose that we are seeking the solution of

$$\frac{dy}{dx} = f(x, y)$$

with $y = y_0$ at $x = x_0$. After n applications of the method we will have reached the point (x_n, y_n) on the solution – or rather, an approximation to the solution. The $(n + 1)$th step is as follows.

A first estimate of $y_{n+1} - y_n$ is given by Euler's method. Call this estimate k_1:

$$k_1 = hf(x_n, y_n) \tag{5.36}$$

where, as usual, $h = x_{n+1} - x_n$.

Thus, the point $(x_n + h, y_n + k_1)$ is a *first approximation* to the next point on the solution. It is based on the proposition that the average slope (dy/dx) of the solution curve over the interval from x_n to x_{n+1} is the same as that at the beginning of the interval, viz. $f(x_n, y_n)$. This is, in fact, the simple Euler method. However, we are justified in supposing that if the slope were to be calculated at the *mid-point* of the interval from x_n to x_{n+1}, then the estimate of the change in y which would be predicted by the use of this average slope would be better than that given by the use of the slope at the start of the interval. The value of x at the mid-point is simply $x_n + h/2$, and our best estimate (indeed, it is our *only* estimate at this stage) of the corresponding value of y is $y_n + k_1/2$. Accordingly, we next compute

$$k_2 = hf\left(x_n + \frac{h}{2}, y_n + \frac{k_1}{2}\right) \tag{5.37}$$

which yields $(x_n + h, y_n + k_2)$ as a better approximation to the next point along the solution. Therefore, extending the above argument, $(x_n + h/2, y_n + k_2/2)$ should be a still better approximation to the location of the mid-point of the interval, and if we now calculate

$$k_3 = hf\left(x_n + \frac{h}{2}, y_n + \frac{k_2}{2}\right) \tag{5.38}$$

then $(x_n + h, y_n + k_3)$ will be an even better approximation to the end-point of the interval.

We have now three estimates of the change in y over the interval: k_1, based on the slope dy/dx at the beginning of the interval, and k_2 and k_3 based on the slopes at two approximations to the mid-point of the interval. A fourth estimate is found by calculating the slope at the end of the interval (or rather, at the latest estimate of the end of the interval):

$$k_4 = hf(x_n + h, y_n + k_3) \tag{5.39}$$

The value that we finally accept for the change in y is a weighted average of these four estimates:

$$y_{n+1} = y_n + \tfrac{1}{6}(k_1 + 2k_2 + 2k_3 + k_4) \tag{5.40}$$

Notice that more weight is given to k_2 and k_3, which were calculated at (approximations to) the mid-point of the interval, than to k_1 and k_4, which were found at the two ends of the interval.

The sequence of steps (5.36)–(5.40) constitutes a Runge–Kutta method. In theory, there are many methods of this general class; in practice, only a few are used – and this is probably the commonest of them.

For example, consider the equation

$$\frac{dy}{dx} = x + y$$

with $y = 1$ at $x = 1$. Choose $h = 0.1$. Then we calculate the value of y at the end of the first interval, i.e. at $x = 1.1$, as follows:

$$k_1 = 0.1(1.0 + 1.0) = 0.2$$
$$k_2 = 0.1(1.05 + 1.1) = 0.215$$
$$k_3 = 0.1(1.05 + 1.1075) = 0.2157$$
$$k_4 = 0.1(1.1 + 1.2157) = 0.2316$$

(to four significant figures). Hence

$$y(1.1) = 1.0 + \tfrac{1}{6}\{0.2 + 2(0.215) + 2(0.2157) + 0.2316\}$$
$$= 1.2155$$

This completes one step of the solution. The new value of y, corresponding to $x = 1.1$, is 1.2155. The solution then continues

$$k_1 = 0.1(1.1 + 1.2155) = 0.2316$$
$$k_2 = 0.1(1.15 + 1.3313) = 0.2481, \text{ etc.}$$

The error incurred by the use of this Runge–Kutta method cannot be readily estimated. However, it is known to be of order h^5, and thus has an accuracy which is comparable with that of Milne's method.

A word of warning to Fortran programmers is warranted. In the absence of explicit or implicit type statements Fortran assumes that all variables with names beginning with I, J, K, L, M or N are integer variables. However, the symbol 'k' for the estimates of $(y_{n+1} - y_n)$ is firmly entrenched in the literature, and to change it here may cause confusion to those reading other texts. A common error made by students writing their first R–K program is to use the Fortran symbols K1, K2, etc., for these same quantities, but to forget to declare them to be real variables. If the Ks are not declared to be real, then their values will be truncated to integers and, of course, the program will give the wrong results.

5.10 Runge–Kutta–Merson method

There is a class of Runge–Kutta methods which does permit an error estimate to be found. It is exemplified by Merson's modification* (R–K–M), described by the following equations:

$$k_1 = hf(x_n, y_n)$$
$$k_2 = hf\left(x_n + \frac{h}{3}, y_n + \frac{k_1}{3}\right)$$
$$k_3 = hf\left(x_n + \frac{h}{2}, y_n + \frac{k_1 + k_2}{6}\right)$$
$$k_4 = hf\left(x_n + \frac{h}{2}, y_n + \frac{k_1 + 3k_3}{6}\right)$$
$$k_5 = hf\left(x_n + h, y_n + \frac{k_1 - 3k_3 + 4k_4}{2}\right)$$
$$y_{n+1} = y_n + \tfrac{1}{6}(k_1 + 4k_4 + k_5) + O(h^5)$$

Notice that five values of k are needed, i.e. there is 25% more work involved in the use of the R–K–M method than for the R–K method. However, these values of k allow an estimate of the local truncation error (i.e. the error of the current step) to be found. It is

$$\varepsilon = \tfrac{1}{15}(k_1 - \tfrac{9}{2}k_3 + 4k_4 - \tfrac{1}{2}k_5)$$

This estimate may be compared with the current value of y, and if the relative error is greater than some acceptable figure – say 10^{-4} – then the solution step should be repeated with a smaller value of h. Since the method is of order h^5, halving the step size will reduce the error by a factor of about 32.

On the other hand, if the relative error is much *less* than the acceptable limit, then the solution could continue with an *increased* step size. In theory, if the relative error is less than 1/32 of the tolerance, then the step size could be doubled without causing the relative error to exceed the specified tolerance. In practice, it would be safer to permit h to be doubled only if the error is less than, say, 1/48 of the tolerance. Otherwise, the solution process might hunt with the size being repeatedly doubled and halved.

An algorithm which incorporates adjustment of step size in this fashion will automatically find the optimum size – i.e. the largest size compatible with the specified local error tolerance. Thus, although R–K–M requires more work *per step* than the regular R–K method, the *total* work for a given solution may well be less, because with R–K a smaller step size than necessary would almost inevitably be chosen, so as to be safe.

---

* Merson, R. H. 1957. An operational method for the study of integration processes. *Proceedings of symposium on data processing*, Weapons Research Establishment, Salisbury, South Australia.

5.11 Application to higher-order equations and to systems

Problems governed by differential equations of the second or higher order can be classified as *initial value problems* or *boundary value problems*. The classification depends on the nature of the auxiliary conditions.

An equation of the nth order requires n auxiliary conditions for its complete solution. Thus the equation

$$y'' + Ay' + By = C \tag{5.41}$$

requires two auxiliary conditions, two pieces of information about y. We may have

$$y = y_1 \quad \text{and} \quad \frac{dy}{dx} = z_1 \quad \text{at } x = x_1 \tag{5.42}$$

for example, or perhaps

$$y = y_1 \quad \text{at } x = x_1 \quad \text{and} \quad y = y_2 \quad \text{at } x = x_2 \tag{5.43}$$

If all of the auxiliary conditions are given at the same value of the independent variable x, we say that we have an *initial value problem*. Thus (5.41) and (5.42) constitute an initial value problem. On the other hand, if the auxiliary conditions are given at two values of x (or conceivably more, for third- or higher-order equations), then we have a *boundary value problem*. Thus (5.41) and (5.43) constitute a boundary value problem.

In this section we will discuss the way in which initial value problems can be solved. Boundary value problems are considered in Section 5.12.

If we define a new variable

$$z = \frac{dy}{dx}$$

then (5.41) can be replaced by the first-order *system*

$$\begin{aligned}\frac{dy}{dx} &= z \\ \frac{dz}{dx} &= -Az - By + C\end{aligned} \tag{5.44}$$

and the initial conditions (5.42) become

$$y = y_1 \quad \text{and} \quad z = z_1 \quad \text{at } x = x_1 \tag{5.45}$$

The set of equations (5.44) is a particular example of a first-order system of ordinary differential equations. The general form of such a system is

$$\begin{aligned}\frac{dy}{dx} &= f_1(x, y, z) \\ \frac{dz}{dx} &= f_2(x, y, z)\end{aligned} \tag{5.46}$$

which requires initial conditions like (5.45) for a solution to be found using a R–K method. An initial value problem can always be written as a system of first-order equations by the introduction of new variables to denote $y^{(i)}$ and, if necessary, $y^{(ii)}$, $y^{(iii)}$, etc.

The Runge–Kutta–Merson procedure can be extended in an obvious manner for such systems. Denoting the five estimates of the change in the value of y over a step h in x by $k_{11}, k_{21}, \ldots, k_{51}$, and the five estimates of the change in the value of z over the same step by $k_{12}, k_{22}, \ldots, k_{52}$, then the solution of (5.46) is given by

$$k_{11} = hf_1(x_n, y_n, z_n)$$
$$k_{12} = hf_2(x_n, y_n, z_n)$$
$$k_{21} = hf_1\left(x_n + \frac{h}{3}, y_n + \frac{k_{11}}{3}, z_n + \frac{k_{12}}{3}\right)$$
$$k_{22} = hf_2\left(x_n + \frac{h}{3}, y_n + \frac{k_{11}}{3}, z_n + \frac{k_{12}}{3}\right)$$
$$k_{31} = hf_1\left(x_n + \frac{h}{3}, y_n + \frac{k_{11} + k_{12}}{6}, z_n + \frac{k_{12} + k_{22}}{6}\right)$$
$$k_{32} = hf_2\left(x_n + \frac{h}{3}, y_n + \frac{k_{11} + k_{12}}{6}, z_n + \frac{k_{12} + k_{22}}{6}\right)$$
$$k_{41} = hf_1\left(x_n + \frac{h}{2}, y_n + \frac{k_{11} + 3k_{31}}{8}, z_n + \frac{k_{12} + 3k_{22}}{8}\right)$$
$$k_{42} = hf_1\left(x_n + \frac{h}{2}, y_n + \frac{k_{11} + 3k_{31}}{8}, z_n + \frac{k_{12} + 3k_{22}}{8}\right)$$
$$k_{51} = hf_1\left(x_n + h, y_n + \frac{k_{11} - 3k_{31} + 4k_{41}}{2}, z_n + \frac{k_{12} - 3k_{32} + 4k_{42}}{2}\right)$$
$$k_{52} = hf_2\left(x_n + h, y_n + \frac{k_{11} - 3k_{31} + 4k_{41}}{2}, z_n + \frac{k_{12} - 3k_{32} + 4k_{42}}{2}\right)$$
$$y_{n+1} = y_n + \tfrac{1}{6}(k_{11} + 4k_{41} + k_{51})$$
$$z_{n+1} = z_n + \tfrac{1}{6}(k_{12} + 4k_{42} + k_{52}) \tag{5.47}$$

with estimates of the local truncation errors in y_{n+1} and z_{n+1} given by

$$\varepsilon_1 = \tfrac{1}{15}(k_{11} - \tfrac{9}{2}k_{31} + 4k_{41} - \tfrac{1}{2}k_{51})$$
$$\varepsilon_2 = \tfrac{1}{15}(k_{12} - \tfrac{9}{2}k_{32} + 4k_{42} - \tfrac{1}{2}k_{52})$$

respectively. If either ε_1 or ε_2 is greater than the specified tolerance, then the whole integration step must be repeated with a smaller value of h.

Notice that the quantities k_{11}, k_{12}, k_{21}, etc., must be calculated in the order shown; for example k_{12} must be known before k_{21} can be found.

Figure 5.4 shows the listing of a Fortran function, MERSON*, which

---

* Extended and adapted for the R–K–M method from a related function RUNGE in Carnahan, B., H. A. Luther & J. O. Wilkes 1969. *Applied numerical methods*. New York: Wiley.

implements the Runge–Kutta–Merson method for a general system of n simultaneous first-order ordinary differential equations. The function is called by a main program, shown in Figure 5.5, which was written to solve the equation

$$\frac{d^2y}{dx^2} + 2\frac{dy}{dx} + 2y = 0$$

$$y = 0 \quad \text{and} \quad \frac{dy}{dx} = 1 \quad \text{at } x = 0$$

or, rather, the equivalent first-order system

$$\frac{dy_1}{dx} = y_2 \qquad \frac{dy_2}{dx} = -2y_1 - 2y_2$$

$$y_1 = 0 \quad \text{and} \quad y_2 = 1 \quad \text{at } x = 0$$

The analytical solution of this problem can be obtained easily (although, of course, this would not be the case when a numerical method was really needed). It is

$$y = e^{-x} \sin x$$

As well as calling the function 'Merson', the main program also calculates the analytical solution and the relative error.

The student should work carefully through these programs, comparing them with equations (5.47). It will be seen that the functions on the right-hand side of (5.46) are evaluated in the main program. The function 'Merson' has two tasks: first to evaluate the arguments of these functions in terms of the 'old' values of y (at the beginning of the step) and the various values of k and, secondly, to adjust the step size as required to keep the local error small enough but not unnecessarily small.

Figure 5.4 The Fortran function MERSON for the solution of a system of differential equations by the Runge–Kutta–Merson process.

```
c
        function merson (n, y, f, x, h)
        real k1, k2, k3, k4, k5
c
c       the dimension statements allow for the solution of up to
c       10 simultaneous first-order differential equations.
c
        dimension k1(10), k2(10), k3(10), k4(10), k5(10),
     +            sumk(10), yold(10), y(n), f(n), eps(10)
c
c
c       epsmax is the relative accuracy tolerance
c
c       ifreq  is the number of x-steps between lines of output
c              (based on the initial step size)
c
c       therefore  h * ifreq  is the maximum step size allowable
c              (otherwise the integration may not finish
c              at the desired maximum value of x)
c
c       icount is the number of steps since the last output
```

Figure 5.4 (*continued*)

```
c
          common epsmax,ifreq,icount
          data m/0/
          hmax = h * ifreq
c
          m = m + 1
          go to (10, 20, 30, 40, 50, 60), m
c
c         first entry to merson
c
   10     merson = 1
          return
c
c         second entry to merson
c
   20     do 25 j=1,n
          yold(j) = y(j)
          k1(j) = h * f(j)
          sumk(j) = k1(j)
   25     y(j) = yold(j) + k1(j) / 3.
c
          x = x + h / 3.
          merson = 1
          return
c
c         third entry to merson
c
   30     do 35 j=1,n
          k2(j) = h * f(j)
   35     y(j) = yold(j) + ( k1(j) + k2(j) ) / 6.
c
          merson = 1
          return
c
c         fourth entry to merson
c
   40     do 45 j=1,n
          k3(j) = h * f(j)
   45     y(j) = yold(j) + ( k1(j) + 3. * k3(j) ) / 8.
c
          x=x + h / 6.
          merson = 1
          return
c
c         fifth entry to merson
c
   50     do 55 j=1,n
          k4(j) = h * f(j)
          sumk(j) = sumk(j) + 4. * k4(j)
   55     y(j) = yold(j) + 0.5 * k1(j) - 1.5 * k3(j) + 2. * k4(j)
c
          x = x + h / 2.
          merson = 1
          return
c
c         sixth and last entry to merson
c
   60     do 65 j=1,n
          k5(j) = h * f(j)
   65     y(j) = yold(j) + ( sumk(j) + k5(j) ) / 6.
          m = 0
          merson = 0
c
c         error check procedure
c
          itest = 0
c
c         calculate the estimate of the errors
c
          do 75 j=1,n
          eps(j) = ( k1(j) - 4.5*k3(j) + 4.*k4(j) - 0.5*k5(j) ) / 15.
```

(*continued*)

Figure 5.4 (*continued*)

```
c
c         bypass the relative error check if y(j) is very small;
c         go instead to 70 to do an absolute error check:
c
              if(abs( y(j) ) .lt. 1.0e-04) go to 70
c
c         if any relative error is greater than epsmax,
c         go immediately to 80 to repeat the step with a
c         new step size of half the previous value:
c
              if(abs( eps(j)/y(j) ) .gt. epsmax) go to 80
c
c         if any relative error is greater than epsmax/48,
c         signal that by making itest different from zero:
c
              if(abs( eps(j) / y(j) ) .gt. epsmax/48.) itest = 1
c
              go to 75
c
c         absolute error check when y(j) is small:
c
   70         if(abs( eps(j) ) .gt. epsmax) go to 80
              if(abs( eps(j) ) .gt. epsmax / 48.) itest = 1
c
   75         continue
c
c         if itest is no longer zero, at least one relative
c         or absolute error is greater than epsmax/48;
c         therefore do not increase the step size:
c
              if(itest .ne. 0) return
c
c         otherwise (i.e., if itest is still zero) go to 95
c         to increase the step size -
c         provided there is not an odd number of steps
c         remaining to the next line of output:
c
              if(mod ( (ifreq-icount-1) , 2) .ne. 0) return
              go to 95
c
c         error is too large; step size is halved:
c
   80         x = x - h
              h = h / 2.
              write (*,85) h
   85         format(' ***** step size reduced to ',f8.6,' *****')
              ifreq = icount + (ifreq - icount) * 2
              do 90 j=1,n
   90         y(j) = yold(j)
              m = 1
              go to 10
c
c         error is too small; step size is doubled:
c
c         (but not if h .ge. hmax already )
c
   95         if ( h .ge. 0.999 * hmax ) return
c
              h= 2. * h
              write (*,100) h
  100         format(' ***** step size increased to ',f8.6,' *****')
              if( ifreq .gt. 1 ) ifreq = 1 + icount + (ifreq-icount) / 2
              return
c
              end
```

Figure 5.5 A Fortran main program to illustrate the solution of a system of ordinary differential equations using the Runge–Kutta–Merson process.

```
c
        common epsmax,ifreq,icount
        dimension f(10),v(10)
c
  200   write (*,210)
  210   format(//,' enter step size, tmax, maximum error and output',/,
       +         '  frequency (in terms of initial step size):',/)
        read (*,*) h,tmax,epsmax,ifreq
c
c       use a non-positive step-size to terminate the program
c
        if( h .le. 0.0 ) stop
c
        range = h * float(ifreq)
        write (*,220)
  220   format(/,' enter number of equations, initial t and',/,
       +         '  initial v(1), v(2), etc.:',/)
        read (*,*) numb,t,(v(j),j=1,numb)
        icount = 0
        write (*,230)
  230   format(//9x,'t             y              y''         true y',/)
c
c       yexact is the analytical solution of the system
c       used in this example - normally it is not known
c       and this part of the program would not appear.
c
  240   yexact = exp(-t) * sin(t)
        write (*,250) t, ( v(j),j=1,numb ), yexact
  250   format(8f14.5)
        ifreq = int (0.5 + range / h)
        if(t .ge. tmax-h/2.) go to 200
  260   k = merson (numb, v, f, t, h)
        if(k .eq. 0) go to 270
c
c       the following statements define the right hand sides
c       of the system of equations being solved:
c
        f(1) = v(2)
        f(2) = -2.* v(1) - 2.* v(2)
        go to 260
  270   icount = icount + 1
        if(icount .ne. ifreq) go to 260
        icount = 0
        go to 240
        end
```

Typical output from this program appears as follows:

```
enter step size, tmax, maximum error and output
frequency (in terms of initial step size):

    .1250    5.0000     .0100     4

enter number of equations, initial t and
initial v(1), v(2), etc.:

    2      .0000      .0000    1.0000
```

(*continued*)

```
         t                y               y'         true y

      .00000           .00000          1.00000        .00000
***** step size increased to  .250000 *****
***** step size increased to  .500000 *****
      .50000           .29079           .24149        .29079
     1.00000           .30952          -.11085        .30956
     1.50000           .22247          -.20676        .22257
     2.00000           .12295          -.17929        .12306
     2.50000           .04904          -.11477        .04913
     3.00000           .00698          -.05622        .00703
     3.50000          -.01060          -.01763       -.01059
     4.00000          -.01385           .00191       -.01386
     4.50000          -.01084           .00852       -.01086
     5.00000          -.00645           .00836       -.00646

enter step size, tmax, maximum error and output
frequency (in terms of initial step size):

     .1250    5.0000    .0001    4

enter number of equations, initial t and
initial v(1), v(2), etc.:

     2    .0000    .0000    1.0000


         t                y               y'         true y

      .00000           .00000          1.00000        .00000
***** step size increased to  .250000 *****
      .50000           .29079           .24149        .29079
***** step size reduced  to  .125000 *****
***** step size increased to  .250000 *****
     1.00000           .30956          -.11080        .30956
     1.50000           .22257          -.20679        .22257
     2.00000           .12305          -.17938        .12306
     2.50000           .04912          -.11488        .04913
     3.00000           .00702          -.05631        .00703
     3.50000          -.01059          -.01768       -.01059
***** step size reduced  to  .125000 *****
     4.00000          -.01386           .00189       -.01386
***** step size increased to  .250000 *****
     4.50000          -.01086           .00852       -.01086
     5.00000          -.00646           .00837       -.00646

enter step size, tmax, maximum error and output
frequency (in terms of initial step size):

     .0000    .0000    .0000    0
```

5.12 Two-point boundary value problems

We have seen that ordinary differential equations of order higher than one can be solved readily provided the requisite number of initial conditions are available. However, problems originating in physics and engineering often require the solution of differential equations in which the data to be satisfied are located at two values of the independent variable.

Consider, for example, the distribution of temperature T along a rod of length L, the ends of which are held at given temperatures T_0 and T_L and

TWO-POINT BOUNDARY VALUE PROBLEMS

which exchanges heat with the surroundings by convection. It can be shown that $T(x)$ is given by the solution of

$$\frac{d}{dx}\left(a\frac{dT}{dx}\right) + bT = c$$

$$T = T_0 \quad \text{at } x = 0 \quad \text{and} \quad T = T_L \quad \text{at } x = L$$

where a, b and c are quantities whose values depend, among other things, on the physical properties of the rod and the surrounding fluid. If these properties are constant (or independent of temperature), then this equation is linear and can be solved analytically. If they are temperature-dependent, then numerical methods may be necessary. In any case, since the boundary conditions on T are prescribed at two values of x, this is a *two-point boundary value problem*.

Another example is provided by the equation describing the buckling of a thin load-bearing column of length L, constrained at each end, the angular displacement ϕ of which is given (under certain conditions) by

$$\frac{d^2\phi}{dx^2} + \lambda \sin \phi = 0$$

$$\phi(0) = \phi(L) = 0$$

where λ may be a constant, or a function of x, or even a function of ϕ.

Among the various techniques available for the solution of such problems, we will consider two: solution by finite differences, and the shooting method.

5.12.1 Finite difference method

This method is particularly suitable for *linear* equations although, as will become apparent, it can also be used for non-linear equations with a considerable increase in computational effort.

We consider the general second-order linear equation

$$P(x)y'' + Q(x)y' + R(x)y + S(x) = 0 \tag{5.48}$$

with boundary conditions

$$y = y_1 \quad \text{at } x = x_1 \quad \text{and} \quad y = y_2 \quad \text{at } x = x_2 \tag{5.49}$$

The solution region from x_1 to x_2 is divided into N equal intervals, of size $h = (x_2 - x_1)/N$. The end-points of each interval are called *mesh points* (the names *grid point* and *node* are also used), and are numbered from 1 to $(N + 1)$ as shown in Figure 5.6.

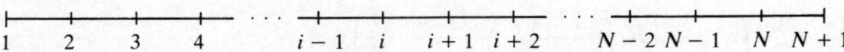

Figure 5.6 The solution mesh.

Values of x at the mesh points are denoted by x_i, and are given by

$$x_i = x_1 + (i - 1)h$$

and the corresponding values of y are denoted by y_i.

The derivatives in (5.48) are replaced by their finite difference approximations. Using central differences,

$$y_i'' = \frac{y_{i+1} - 2y_i + y_{i-1}}{h^2} + O(h^2) \qquad (5.50)$$

$$y_i' = \frac{y_{i+1} - y_{i-1}}{2h} + O(h^2) \qquad (5.51)$$

Making the approximation that the truncation errors are negligible, the substitution of (5.50) and (5.51) into (5.48) leads to

$$P_i \frac{y_{i+1} - 2y_i + y_{i-1}}{h^2} + Q_i \frac{y_{i+1} - y_{i-1}}{2h} + R_i y_i + S_i = 0 \qquad (5.52)$$

where P_i, Q_i, R_i and S_i denote $P(x_i)$, $Q(x_i)$, $R(x_i)$ and $S(x_i)$, respectively. Rearranging (5.52), we obtain

$$\left(\frac{P_i}{h^2} - \frac{Q_i}{2h}\right)y_{i-1} + \left(R_i - \frac{2P_i}{h^2}\right)y_i + \left(\frac{P_i}{h^2} + \frac{Q_i}{2h}\right)y_{i+1} = -S_i \qquad (5.53)$$

An equation like (5.53) can be constructed for each point at which y is unknown, i.e. for the *internal* mesh points, those for which $i = 2, 3, \ldots, N$. There will be $N - 1$ equations for the $N - 1$ unknown quantities y_2, $y_3, \ldots, y_N$.

The values y_1 and y_{N+1} are known: they are given by the boundary conditions (5.49). They appear in the first equation ($i = 2$) and the last equation ($i = N$), respectively. When $i = 2$, y_1 appears in (5.53); since it is known, that term is transferred to the right-hand side of the equation. Similarly, when $i = N$ the term containing y_{N+1} is written on the right-hand side. Thus, we obtain the system of algebraic equations:

$$\left(R_2 - \frac{2P_2}{h^2}\right)y_2 + \left(\frac{P_2}{h^2} + \frac{Q_2}{2h}\right)y_3 \qquad \qquad = -S_2 - \left(\frac{P_2}{h^2} - \frac{Q_2}{2h}\right)y_1$$

$$\left(\frac{P_i}{h^2} - \frac{Q_i}{2h}\right)y_{i-1} + \left(R_i - \frac{2P_i}{h^2}\right)y_i + \left(\frac{P_i}{h^2} + \frac{Q_i}{2h}\right)y_{i+1} = -S_i$$

$$i = 3, 4, \ldots, (N - 1)$$

TWO-POINT BOUNDARY VALUE PROBLEMS

$$\left(\frac{P_N}{h^2} - \frac{Q_N}{2h}\right) y_{N-1} + \left(R_N - \frac{2P_N}{h^2}\right) y_N = -S_N - \left(\frac{P_N}{h^2} + \frac{Q_N}{2h}\right) y_{N+1} \quad (5.54)$$

the solution of which is an approximation to the solution of (5.48).

This system of $(N - 1)$ equations is tridiagonal. It may be written

$$\begin{bmatrix} b_2 & c_2 & & & & & \\ a_3 & b_3 & c_3 & & & & \\ & a_4 & b_4 & c_4 & & & \\ & & & \vdots & & & \\ & & & a_i & b_i & c_i & \\ & & & & & \vdots & \\ & & & & & a_N & b_N \end{bmatrix} \begin{bmatrix} y_2 \\ y_3 \\ y_4 \\ \vdots \\ y_i \\ \vdots \\ y_N \end{bmatrix} = \begin{bmatrix} d_2 \\ d_3 \\ d_4 \\ \vdots \\ d_i \\ \vdots \\ d_N \end{bmatrix}$$

where

$$a_2 = 0$$

$$a_i = \frac{P_i}{h^2} - \frac{Q_i}{2h} \quad i = 3, 4, \ldots, N$$

$$b_i = R_i - \frac{2P_i}{h^2} \quad i = 2, 3, \ldots, N$$

$$c_i = \frac{P_i}{h^2} + \frac{Q_i}{2h} \quad i = 2, 3, \ldots, N - 1$$

$$c_N = 0$$

$$d_2 = -S_2 - \left(\frac{P_2}{h^2} - \frac{Q_2}{2h}\right) y_1$$

$$d_i = -S_i \quad i = 3, 4, \ldots, N - 1$$

$$d_N = -S_N - \left(\frac{P_N}{h^2} - \frac{Q_N}{2h}\right) y_{N+1}$$

Since the system is tridiagonal, it may readily be solved by the Thomas algorithm (Section 3.6).

Figure 5.7 shows the listing of a Fortran main program which calculates the coefficients in terms of the functions P, Q, R and S; these in turn are defined in separate function sub-programs. The solution is found using the subroutine 'Thomas' of Figure 3.2. The example chosen for solution is

$$xy'' - 2y' + 2 = 0 \quad y(0) = y(1) = 0 \quad (5.55)$$

Thus,

$$P(x) = x \quad Q(x) = -2 \quad R(x) = 0 \quad S(x) = 2$$

ORDINARY DIFFERENTIAL EQUATIONS

```
      c
            dimension  x(51), y(51), z(51), a(49), b(49), c(49), d(49)
            common x
      c
      c     initialization
      c
        10 write (*,100)
       100 format (' how many intervals?')
           read (*,*) n
           if (n .le. 0) stop
           np1 = n + 1
           mid = (n + 2)/2
           nm1 = n - 1
           nm2 = n - 2
           write (*,300)
       300 format (' enter xmin,  xmax,  ymin  and  ymax:')
           read (*,*) x(1),x(np1),y(1),y(np1)
       303 format(4g7.2)
           dx = ( x(np1) - x(1) ) / float(n)
           do 20 i = 2,n
        20 x(i) = x(i-1) + dx
      c
      c     calculate the exact solution (not normally known)
      c
           ytrue = x(mid) - (x(mid))**3
      c
      c     calculate coefficients
      c
           b(1) = r(2) - 2.*p(2)/(dx*dx)
           c(1) = p(2) / (dx*dx) + q(2)/(2.*dx)
           d(1) = - s(2) - (p(2) / (dx*dx) - q(2)/(2.*dx))*y(1)
           do 30 i = 2,nm2
           a(i) = p(i+1) / (dx*dx) - q(i+1)/(2.*dx)
           b(i) = r(i+1) - 2.*p(i+1)/(dx*dx)
           c(i) = p(i+1) / (dx*dx) + q(i+1)/(2.*dx)
        30 d(i) = - s(i+1)
           a(nm1) = p(n) / (dx*dx) - q(n)/(2.*dx)
           b(nm1) = r(n) - 2.*p(n)/(dx*dx)
           d(nm1) = - s(n) - (p(n) / (dx*dx) - q(n)/(2.*dx))*y(np1)
      c
           call thomas (a, b, c, d, z, nm1)
      c
           do 40 i = 2,n
        40 y(i) = z(i-1)
      c
           write (*,500) (x(i), i = 1,np1)
       500 format('  x: ',9f8.4)
           write(*,600) (y(i), i = 1,np1)
       600 format('  y: ',9f8.4)
      c
      c     calculation of error at the mid-point
      c
           error = ytrue - y(mid)
           write (*,700)  ytrue, error
       700 format (/,' true mid-point value is ',f8.4,
          +        ';  the error there is ',f6.4,/)
           go to 10
           end

           function p(n)              function r(n)
           common x(49)               common x(49)
           p = x(n)                   r = 0.0
           return                     return
           end                        end

           function q(n)              function s(n)
           common x(49)               common x(49)
           q = - 2.                   s = 2.0
           return                     return
           end                        end
```

Figure 5.7 A program for the solution of a two-point boundary value problem by finite differences.

Typical output from this program appears as follows:

```
how many intervals?
        4
enter xmin,  xmax,   ymin  and    ymax:
        .00    1.0    .00          .00
   x:  .0000  .2500  .5000  .7500 1.0000
   y:  .0000  .2500  .4000  .3500  .0000

true mid-point value is    .3750; the error there is -.0250

how many intervals?
        8
enter xmin,  xmax,   ymin  and    ymax:
        .00    1.0    .00          .00
   x:  .0000  .1250  .2500  .3750  .5000  .6250  .7500  .8750 1.0000
   y:  .0000  .1250  .2381  .3274  .3810  .3869  .3333  .2083  .0000

true mid-point value is    .3750; the error there is -.0060

how many intervals?
        0
```

The analytical solution of (5.55) is $y = x - x^2$. Hence, we can study the accuracy of the numerical solution as a function of h, the mesh size. Figure 5.8 shows the error in the computed mid-point value, the true value of which is 0.375. It can be seen that the error reduces as the mesh size h is reduced, which is clearly a desirable feature of the solution process. In fact, the error is very closely proportional to h^2, a result which is consistent with the order of the truncation errors in (5.50) and (5.51).

Using this solution technique, we do not have the concepts of *local* and *global* truncation errors. Since all $N - 1$ finite difference approximations are solved simultaneously, the inheritance of error from one mesh point to the next does not occur as it does when the approximations are solved consecutively. Thus (5.53) is $O(h^2)$ and we expect the solution also to be $O(h^2)$.

The fact that the numerical solution approaches the analytical solution as $h \to 0$ is described by saying that the numerical procedure is *convergent**.

* Remember that the term 'convergence' was also introduced in Chapter 2 to describe an iterative process which approached a limit as the iterations progressed.

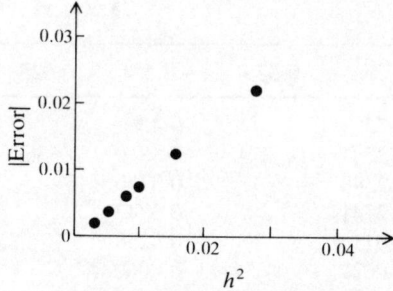

Figure 5.8 Quadratic convergence.

Since the truncation error of (5.53) is $O(h^2)$, we say that the equation is *quadratically convergent*.

Not all processes are convergent – in either sense of the word. One could imagine an iterative procedure being used to solve the non-linear finite difference approximations resulting from a non-linear two-point boundary value problem. Then the process could be *convergent* in the sense that successive estimates of the solution approach a limit as the number of iterations tends to infinity, but – conceivably – *divergent* in the sense that *that limit* does not approach the *true solution* as the mesh size tends to zero.

It is often not easy to demonstrate that a given procedure is convergent (in the present meaning of the term), and we shall not discuss how this might be done. It can be assumed that the methods discussed here are convergent.

5.12.2 Richardson's extrapolation

The concept of Richardson's extrapolation, introduced in Chapter 4 in connection with interpolation and the numerical integration and differentiation of functions, can also be applied to the finite difference solution of a differential equation if the order of the truncation error is known. The technique follows that developed in Section 4.16. It is necessary to perform the solution procedure for two different values of h; values which differ by a factor of two are usually chosen. The two solutions thus obtained will have some mesh points in common, and the computed values at those points can be used to generate an *extrapolated solution* which (subject to the approximation involved) will be the solution which would be achieved with a mesh size of zero.

The first three rows of Table 5.2 show the values of x and the corresponding values of y_1 and y_2 – the numerical solutions of (5.55) – for $h_1 = 0.25$ and $h_2 = 0.125$. There are mesh points in common to the two solutions at $x = 0.25$, 0.5 and 0.75. (The boundary points, which are also

Table 5.2 The use of Richardson's extrapolation to improve the accuracy of the finite difference solution of a differential equation.

				x			
	0.125	0.25	0.375	0.5	0.625	0.75	0.875
y_1		0.25		0.4		0.35	
y_2	0.1250	0.2381	0.3274	0.3810	0.3869	0.3333	0.2083
y_t		0.2344		0.3750		0.3281	
y_e		0.2341		0.3747		0.3277	
e_1		−0.0156		−0.0250		−0.0219	
e_2		−0.0037		−0.0060		−0.0052	
e_e		0.0003		0.0003		0.0004	

common to the two meshes, have values of y which are given, and therefore do not have to be computed. They have been omitted from the table.)

The fourth row in the table shows y_t, the true solution at the three common mesh points, computed from the known analytical solution.

Because $h_2 = h_1/2$, and because the truncation error is $O(h^2)$, Richardson's extrapolation (4.59) becomes

$$y_e = y_2 + (y_2 - y_1)/3$$

This formula has been used to compute the values in the fifth row of Table 5.2.

The last three rows of the table show the errors e_1 and e_2 in the two computed values and the error e_e in the extrapolated values. It can be seen first, by comparing e_1 and e_2, that the error has been reduced by a factor of approximately four by halving the step size. The error reduction is not exactly fourfold because the step sizes used are fairly large; the error is only really proportional to h^2 in the limit $h \to 0$. However, the reduction is almost fourfold. Secondly, the errors in the extrapolated values are very much smaller even than those obtained using h_2; they are more than an order of magnitude smaller. With very little effort, therefore, a high quality solution has been obtained.

5.12.3 *Derivative boundary conditions*

Occasionally, the boundary conditions differ from those given in (5.49). The most frequent alternative is

$$y = y_1 \quad \text{at } x = x_1 \quad \text{and} \quad \frac{dy}{dx} = z_2 \quad \text{at } x = x_2 \qquad (5.56)$$

i.e. the right-hand boundary condition is a *derivative* condition. In this case, y_{N+1} is not known, and must be calculated as part of the solution.

There are two ways of doing this. In the first, we approximate dy/dx at x_2 by the backward difference approximation

$$\left.\frac{dy}{dx}\right|_{N+1} = \frac{y_{N+1} - y_N}{h} + O(h) \qquad (5.57)$$

Using the boundary condition (5.56), we find that

$$y_{N+1} = y_N + h z_2 \qquad (5.58)$$

Equation (5.58) must be appended to the system (5.54), the last of which must be modified since y_{N+1} is not now known. The last two equations of the new system become

$$\left(\frac{P_N}{h^2} - \frac{Q_N}{2h}\right)y_{N-1} + \left(R_N - \frac{2P_N}{h^2}\right)y_N + \left(\frac{P_N}{h^2} + \frac{Q_N}{2h}\right)y_{N+1} = -S_N \qquad (5.59)$$

and
$$-y_N + y_{N+1} = hz_2 \qquad (5.60)$$

The system is not now tridiagonal, but can be made so by using (5.60) to eliminate y_{N+1} from (5.59), yielding

$$\left(\frac{P_N}{h^2} - \frac{Q_N}{2h}\right)y_{N-1} + \left(R_N - \frac{P_N}{h^2} + \frac{Q_N}{2h}\right)y_N = -S_N - hz_2\left(\frac{P_N}{h^2} + \frac{Q_N}{2h}\right) \qquad (5.61)$$

The system (5.54), with (5.61) replacing the last equation of the system, is solved for $y_2, y_3, \ldots, y_N$. Then (5.58) is used to find y_{N+1}.

This procedure, although straightforward, suffers from the defect that the approximation (5.57) to the boundary condition, being only of order h, is less accurate than the approximation used for the differential equation itself, which is of order h^2. The increased error will not be confined to the value of y_{N+1}, but will contaminate all computed values of y.

A preferable alternative is to use a central difference approximation to dy/dx at x_2. This requires the introduction of a hypothetical mesh point at a location x_{N+2}, beyond the end of the solution region, as shown in Figure 5.9.

$$- - -\!\!\!+\!\!\!-\!\!\!-\!\!\!+\!\!\!-\!\!\!-\!\!\!+\!\!\!-\!\!\!-\!\!\!+\cdots\vdots$$
$$\quad N-2\; N-1\quad N\quad N+1\; N+2$$

Figure 5.9 Hypothetical mesh point for derivative boundary condition.

If this new mesh point existed, it would have associated with it a value of y of y_{N+2}. In terms of y_{N+2} we may write

$$\left.\frac{dy}{dx}\right|_{N+1} = \frac{y_{N+2} - y_N}{2h} + O(h^2) \qquad (5.62)$$

or

$$y_{N+2} = y_N + 2hz_2 \qquad (5.63)$$

When (5.53) is applied at the point $i = N + 1$, the value of y_{N+2} is required. This is supplied by (5.63), and we obtain

$$\frac{2P_{N+1}}{h^2}y_N + \left(R_{N+1} - \frac{2P_{N+1}}{h^2}\right)y_{N+1} = -S_{N+1} - 2hz_2\left(\frac{P_{N+1}}{h^2} + \frac{Q_{N+1}}{2h}\right) \qquad (5.64)$$

The system comprising (5.54) together with (5.64) is then solved for $y_2, y_3, \ldots, y_{N+1}$ using the Thomas algorithm. The error in (5.63) is $O(h^2)$, compatible with the finite difference approximation to the differential equation itself.

5.12.4 The shooting method

An alternative procedure for solving a two-point boundary value problem involves its conversion to an initial value problem by the determination of sufficient additional condition(s) at one boundary. A second-order equation will require two initial conditions, and only one is provided; a third-order equation will require three, and only one or two are given; and so on. The missing initial conditions are determined in a way which causes the given conditions at the other boundary to be satisfied.

In summary, the steps involved are:

(1) split the second- (or higher-) order equation into two (or more) equivalent first-order equations as described in Section 5.11;
(2) estimate values for the missing initial condition or conditions;
(3) integrate the equations as an initial value problem;
(4) compare the solution at the final boundary with the given final boundary condition(s); if they do not agree, then
(5) adjust the estimated values of the missing initial condition(s); and
(6) repeat the integration until the process converges.

On the assumption that the final conditions are continuously related to the assumed initial conditions*, the adjustment of the assumed values can be done systematically. If the dependence is, in addition, monotonic, then the adjustment is relatively simple. If it is not, then the adjustment can still be made, but may require a little care.

As an illustration, consider the equation

$$y'' + 2xy' - 6y - 2x = 0$$
$$y = 0 \quad \text{at } x = 0 \quad \text{and} \quad y = 2 \quad \text{at } x = 1 \quad (5.65)$$

This is a two-point boundary value problem. Its solution happens to be

$$y = x^3 + x \quad (5.66)$$

which will allow us to check on the accuracy of our numerical solution. It could be solved by the finite difference method, but instead we will convert it numerically into an initial value problem.

The differential equation in (5.65) is replaced by

$$y' = z \quad z' = -2xz + 6y + 2x \quad (5.67)$$

We have one initial condition for (5.67): the value of y at $x = 0$. Since we do not know the value of z (i.e. y') at $x = 0$, we must assume a value: we may be able to make an estimate from our knowledge of the original problem

---

* If this assumption is not valid, then the problem is not well-posed and cannot be solved in this way.

from which the differential equation derived; otherwise we might be forced simply to make a guess. We write

$$y'(0) = \alpha$$

where α is the chosen numerical value. We are now able to do the integration of (5.65) between $x = 0$ and $x = 1$. We note in passing that for the particular example chosen, (5.66) shows that the correct value for α is 1.

Whatever value we choose first for α is not likely to be correct, so the solution we compute will not be correct. In particular, the right-hand boundary condition ($y = 2$ at $x = 1$, in this case) will not be satisfied. We now try to adjust α to get that condition to be satisfied. If the symbol Y is used to denote $y(1)$, and recognizing that Y is a function of α, then what we seek to do is to solve the equation $Y(\alpha) = 2$. In general, using the notation of (5.49), we have to solve

$$Y(\alpha) = y_2 \tag{5.68}$$

The analytical nature of the function $Y(\alpha)$ is not known, but we can compute values for it numerically as described: we give α various values, and perform the integration by a Runge–Kutta method, for example.

Figure 5.10 shows the results which are obtained for some particular values of α. When $\alpha = 0.5$, the value of Y is too small; when $\alpha = 2$, it is too large; and when $\alpha = 1$, Y is exactly 2, which is the correct value. The shooting method derives its name from the fact that, by adjusting α, we are shooting at the 'target' $Y = y(1) = y_2$.

To solve (5.68), we can use the Newton–Raphson method. Since $Y(\alpha)$ is not known analytically, we have to calculate the derivative $Y'(\alpha)$ numerically:

$$Y'(\alpha) \approx \frac{Y(\alpha + \delta\alpha) - Y(\alpha)}{\delta\alpha} \tag{5.69}$$

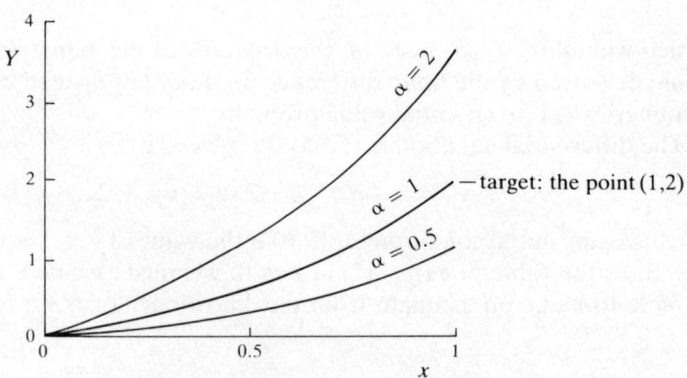

Figure 5.10 The shooting method.

In other words, the integration must be performed twice with two slightly different values of α to allow a numerical differentiation to be made. An improved value of α may then be found from

$$\alpha_{\text{new}} = \alpha - \frac{Y(\alpha) - y_2}{Y'(\alpha)}$$

This process is continued until $Y(\alpha)$ is 'sufficiently' close to y_2.

The present example (5.65) is a linear equation, and $Y(\alpha)$ is therefore a linear function; hence it should converge immediately. Two iterations might, in fact, be necessary because of the approximation involved in (5.69). Table 5.3 shows results from a program written to solve a two-point boundary value problem in this way. The integration was performed using the R–K–M method (Fig. 5.4). The convergence criterion 'epsmax' was set at 0.0001; this criterion was used both for the automatic adjustment of step

Table 5.3 An illustration of the shooting method for $y'' + 2xy' - 6y - 2x = 0$ with $y(0) = 0$ and $y(1) = 2$.

Initial $\alpha = 2.00000$

x	y	y'	True y ($x^3 + x$)
0.00000	0.00000	2.00000	0.00000
0.20000	0.41333	2.20000	0.20800
0.40000	0.90667	2.80000	0.46400
0.60000	1.56000	3.80000	0.81600
0.80000	2.45333	5.20000	1.31200
1.00000	3.66666	7.00000	2.00000

End of first pass. Continue with $\alpha = 2.02000$

0.00000	0.00000	2.02000	0.00000
0.20000	0.41744	2.22160	0.20800
0.40000	0.91552	2.82640	0.46400
0.60000	1.57488	3.83440	0.81600
0.80000	2.47616	5.24560	1.31200
1.00000	3.70000	7.06000	2.00000

End of second pass. Repeat with $\alpha = 0.99997$

0.00000	0.00000	0.99997	0.00000
0.20000	0.20799	1.11997	0.20800
0.40000	0.46399	1.47996	0.46400
0.60000	0.81598	2.07995	0.81600
0.80000	1.31197	2.91993	1.31200
1.00000	1.99995	3.99991	2.00000

Shooting procedure has converged. Stop.

size of the R–K–M procedure and for the solution of (5.68). The step size $\delta\alpha$ in (5.69) was chosen to be 0.01α. The initial value selected for α was 2; the correct value, within the specified tolerance, was found in one iteration.

5.13 Non-linear two-point boundary value problems

If the differential equation to be solved is non-linear, then it is apparent that so, too, will be its finite difference approximation. Hence, solution of such a problem by finite difference methods will require a non-linear system of equations to be solved. Almost invariably, iterative techniques must be used.

For example, the general second-order equation

$$y'' = f(x, y, y')$$

can be approximated by

$$\frac{y_{i+1} - 2y_i + y_{i-1}}{h^2} = f\left(x_i, y_i, \frac{y_{i+1} - y_{i-1}}{2h}\right) \quad (5.70)$$

at the nodes of a uniform mesh along the x-axis.

The solution of (5.70) can be tackled in two ways. The simplest is to rewrite it in the form

$$y_i = \tfrac{1}{2}(y_{i-1} + y_{i+1}) - \frac{h^2}{2} f\left(x_i, y_i, \frac{y_{i+1} - y_{i-1}}{2h}\right)$$

which suggests the obvious iteration scheme

$$y_i^{(n+1)} = \tfrac{1}{2}(y_{i-1}^{(n+1)} + y_{i+1}^{(n)}) - \frac{h^2}{2} f\left(x_i, y_i^{(n)}, \frac{y_{i+1}^{(n)} - y_{i-1}^{(n+1)}}{2h}\right) \quad (5.71)$$

where the superscript (n) denotes an iteration number. It has been assumed that the calculations are made in ascending order of mesh point number i, so that an improved value of y_{i-1} will have been found before the next value of y_i is calculated and is used in that calculation. In other words, Gauss–Seidel iteration is used, not Jacobi iteration (see Section 3.8).

This process is quite straightforward to implement. If the boundary conditions are like (5.49), in which the values of y are prescribed at each end of the solution region (i.e. at mesh points 1 and $N + 1$), then only the internal mesh point values of y have to be calculated. Starting with some assumed set of values for y, (5.71) is applied point by point, i.e. for $i = 2$, $3, \ldots, N$, to yield a new distribution. The process is repeated until two successive estimates differ for all values of i by less than some prescribed amount.

If one of the boundary conditions is a derivative condition like (5.56), it can be handled as described in the previous section.

Unfortunately, the convergence of this scheme cannot be guaranteed. However, if a parameter σ is chosen to satisfy

$$\sigma \geq \frac{h^2}{2} \max \left| \frac{\partial f(x, y, y')}{\partial y} \right| \quad (5.72)$$

over the solution region, then the scheme

$$(1 + \sigma)y_i^{(n+1)} = \tfrac{1}{2}(y_{i-1}^{(n+1)} + y_{i+1}^{(n)}) + \sigma y_i^{(n)} - \frac{h^2}{2} f\left(x_i, y_i^{(n)}, \frac{y_{i+1}^{(n)} - y_{i-1}^{(n+1)}}{2h}\right) \quad (5.73)$$

can be shown to be convergent. The quantity σ provides a measure of damping between the 'old' value $y^{(n)}$ and the 'new' value $y^{(n+1)}$. A fraction $\sigma/(\sigma + 1)$ of the old value is added to a fraction $1/(\sigma + 1)$ of the value given by (5.71). For positive values of σ, under-relaxation is applied, while negative values of σ correspond to over-relaxation.

The only problem is the determination of σ. In general, it will not be possible to evaluate the partial derivative required in (5.72) since y is not known over the solution region. However, since (5.72) imposes only a minimum value on σ, σ can always be chosen sufficiently large to ensure that the condition is satisfied. Of course, if σ is very large, then it can be seen from (5.73) that the process will be slow to converge. However, it will converge eventually, and a judicious adjustment of σ as the iterations progress will enable the optimum value to be found empirically.

On the other hand, over-relaxation – a negative value for σ – may increase the rate of convergence of (5.73) over that of (5.71).

An alternative procedure is to apply iteration to (5.70) in the form

$$\frac{y_{i+1}^{(n+1)} - 2y_i^{(n+1)} + y_{i-1}^{(n+1)}}{h^2} = f\left(x_i, y_i, \frac{y_{i+1}^{(n)} - y_{i-1}^{(n)}}{2h}\right) \quad (5.74)$$

This is a tridiagonal system. There are three unknowns in each equation, and the Thomas algorithm may be used to obtain the solution. Again, an initial estimate of y is required.

Non-linear boundary value problems may also be solved by the shooting method. Indeed, the same program may be used for linear and non-linear problems; it is necessary only to change the statements defining the 'right-hand sides' of the first-order system into which the equation is split.

Worked examples

1. Solve

$$\frac{dy}{dx} = 2 \cos x + y \quad \text{with} \quad y = -1 \quad \text{at } x = 0$$

using a Taylor series.

200   ORDINARY DIFFERENTIAL EQUATIONS

$$y_{n+1} = y_n + hy_n^{(i)} = \frac{h^2}{2!} y_n^{(ii)} + \frac{h^3}{3!} y_n^{(iii)} + \frac{h^4}{4!} y_n^{(iv)} + \cdots$$

where $y_n^{(i)} = 2 \cos x_n + y_n$
$y_n^{(ii)} = -2 \sin x_n + 2 \cos x_n + y_n$
$y_n^{(iii)} = -2 \sin x_n + y_n$
$y_n^{(iv)} = y_n$
$y_n^{(v)} = 2 \cos x_n + y_n$, etc.

The solution is found by selecting h and substituting into the series. With $h = 0.5$, the first few steps of the solution are:

n	x_n	y_n	$y_n^{(i)}$	$y_n^{(ii)}$	$y_n^{(iii)}$	$y_n^{(iv)}$	$y_n^{(v)}$	y_{n+1}
1	0	-1	1	1	-1	-1	1	-0.3982
2	0.5	-0.3982	1.3570	0.3981	-1.3570	-0.3982	1.3570	0.3011
3	1	0.3011	1.3817	-0.3012	-1.3818	0.3011	1.3817	0.9267
4	1.5	0.9267	1.0682	-0.9268	-1.0683	0.9267	1.0682	1.3254
5	2	1.3254	...					

2. What is the largest step size that can be used with the modified Euler method to solve

$$\frac{dy}{dx} = x + y \quad \text{with } y = 1 \quad \text{at } x = 0?$$

The modified Euler corrector is

$$y_{n+1} = y_n + \frac{h}{2} \{f(x_n, y_n) + f(x_{n+1}, y_{n+1})\}$$

$$= y_n + \frac{h}{2} (x_n + y_n + x_{n+1} + y_{n+1})$$

$$= F(y_{n+1}) \quad \text{say}$$

Convergence requires $|F'(y_{n+1})| < 1$. Since $F'(y_{n+1}) = h/2$, the step size limit is 2.

The results of the first few iterations, for various values of h, are shown in Table 5.4. For this linear problem, the iteration equation can be solved analytically. The solution, for the given initial condition, is

$$y_{n+1} = (2 + h + h^2)/(2 - h)$$

Using $h = 1$, the iterations converge to $y_{n+1} = 4$, which is the solution of the corrector equation (but *not*, because of the truncation error in the corrector, the solution of the differential equation). The error – the difference between each value of y_{n+1} and 4.000 – reduces by a factor of 0.5 every iteration, in

WORKED EXAMPLES

Table 5.4 Modified Euler solution for Worked Example 2.

h	1	1.9	2	2.1
x_n	0	0	0	0
y_n	1	1	1	1
x_{n+1}	1	1.9	2	2.1
y_{n+1}^p	3	6.51	7	7.51
$y_{n+1}^{c,1}$	3.5	9.940	11	12.141
$y_{n+1}^{c,2}$	3.75	13.198	15	17.003
$y_{n+1}^{c,3}$	3.875	16.293	19	22.108
$y_{n+1}^{c,4}$	3.938	19.233	23	27.468
$y_{n+1}^{c,5}$	3.969	22.026	27	33.096
$y_{n+1}^{c,6}$	3.984	27.201	31	39.006
'Solution'	4.000	75.1		−85.1

accordance with the value of $F'(y_{n+1})$. The solution of the differential equation at $x = 1$ is 3.437; the numerical value is thus about 16% high.

Using $h = 1.9$, the iterations converge very slowly to the analytical solution of the corrector equation, which is 75.1. The error reduces by a factor of $F' = 0.95$ every iteration. However, the solution of the differential equation at $x = 1.9$ is 10.472. The numerical value is thus completely wrong.

With $h = 2$, the iterations do not converge; the change in y_{n+1} from one iteration to the next is constant. The corrector equation has no (finite) solution, although the differential equation, of course, does. With $h = 2.1$, the solution diverges; the successive changes in y_{n+1} are growing in size. The corrector equation has a solution (-85.1) which has no relationship to the differential equation solution (13.232) at $x = 2.1$.

We therefore see that the theoretical limit on h is verified. We also see that it is most undesirable to use a value of h close to that limit; the iterations converge very slowly, and they converge to a very wrong value!

3. Solve the equation

$$y' = \cos x - \sin x + y = f(x, y)$$

with $y = 0$ at $x = 0$, using Milne's method. Obtain starting values by the modified Euler method. Compute the solution with several step sizes, and try to obtain an accurate value for $y(10)$.

We will use, for Milne's method, the step sizes: $h_M = 0.2$, 0.1 and 0.05 and, following the argument in Section 5.7, use $h_E = 0.01, 0.0025$ and 0.001, respectively. As a result of using such small step sizes, the output is very extensive and only selected values are listed here.

The starting values calculated using the modified Euler method and $h_E = 0.01$ are given in Table 5.5.

Table 5.5 Fine mesh starting values for Worked Example 3.

x	y	x	y	x	y
0.0000	0.0000				
0.0100	0.0100	0.2100	0.2085	0.4100	0.3986
0.0200	0.0200	0.2200	0.2182	0.4200	0.4078
⋮	⋮	⋮	⋮	⋮	⋮
0.0500	0.0500	0.2500	0.2474	0.4500	0.4350
⋮	⋮	⋮	⋮	⋮	⋮
0.1900	0.1889	0.3900	0.3802	0.5900	0.5564
0.2000	0.1987	0.4000	0.3894	0.6000	0.5646

The values at the foot of each column are those which are needed to continue by Milne's method. All the other values – and all of the effort that went into obtaining them – are discarded.

Table 5.6 shows (a portion of) the rest of the solution – the predicted value, the final converged value of the corrector equation and the analytical solution. It can be seen that the numerical solution gradually deviates from the true value; the error at $x = 10$ is 0.0469.

Table 5.6 Continuation of the solution of Worked Example 3.

x	y_{pred}	y_{corr}	Exact
0.0000		0.0000	0.0000
0.2000		0.1987	0.1987
0.4000		0.3894	0.3894
0.6000		0.5646	0.5646
0.8000	0.7173	0.7174	0.7174
1.0000	0.8414	0.8415	0.8415
1.2000	0.9320	0.9320	0.9320
1.4000	0.9854	0.9854	0.9854
1.6000	0.9995	0.9996	0.9996
1.8000	0.9738	0.9738	0.9738
2.0000	0.9093	0.9093	0.9093
⋮	⋮	⋮	⋮
3.0000	0.1412	0.1411	0.1411
⋮	⋮	⋮	⋮
9.0000	0.3949	0.3949	0.4121
⋮	⋮	⋮	⋮
10.0000	−0.5908	−0.5909	−0.5440

WORKED EXAMPLES

The results at $x = 10$ of performing the calculations with three different step sizes are:

h_E	h_M	y_{corr}	Error
0.0100	0.20	−0.5909	0.0469
0.0025	0.10	−0.5423	−0.0017
0.0010	0.05	−0.5441	0.0001

The discrepancies between the numerical and true values are diminishing, although not monotonically.

In a real situation, where the solution is not known, we could not compute the errors. Can we use Richardson's extrapolation? Not in the usual manner, because we cannot be sure what the order of the global truncation error is. It is true that we know the local errors of both the modified Euler starting procedure and the Milne continuation, but we cannot evaluate the effects of inherited error from one stage of the calculation to the next, or of the residual errors in the iterative solution of the corrector equation.

However, we can make some progress if we assume that each of the three computed values (call them V_1, V_2 and V_3) are related to the true solution (S) and the respective step size h_i by

$$S - V_i = C h_i^m$$

where C and m are unknown, but are assumed to be constants. The three values of V_i enable us to solve for S, leading to

$$S = \frac{V_2^2 - V_1 V_3}{2V_2 - V_1 - V_3}$$
$$= -0.5440$$

which happens to be the exact value (to the precision stated).

4. Solve the equation

$$y'' = y^2 + y' \text{ with } y(0) = y(1) = 1$$

by the methods of Section 5.13; namely (a) by an iterative method and (b) by the shooting method.

(a) Equation (5.73) becomes

$$(1 + \sigma) y_i^{(n+1)} = \tfrac{1}{2}(y_{i-1}^{(n+1)} + y_{i+1}^n) + \sigma y_i^{(n)} - \frac{h^2}{2}\left((y_i^{(n)})^2 + \frac{y_{i+1}^{(n)} - y_{i-1}^{(n+1)}}{2h}\right)$$

Starting with $y = 1$ for all values of x, and using $\sigma = 0$ and $h = 0.1$, the results of y for the first few iterations are

y(0.0)	y(0.1)	y(0.2)	y(0.3)	y(0.4)	y(0.5)	y(0.6)	y(0.7)	y(0.8)	y(0.9)	y(1.0)
1.000	1.000	1.000	1.000	1.000	1.000	1.000	1.000	1.000	1.000	1.000
1.000	0.995	0.992	0.991	0.990	0.990	0.990	0.990	0.990	0.990	1.000
1.000	0.991	0.986	0.983	0.982	0.981	0.980	0.980	0.979	0.984	1.000
1.000	0.989	0.981	0.977	0.974	0.972	0.971	0.970	0.972	0.980	1.000
1.000	0.986	0.977	0.970	0.966	0.964	0.962	0.962	0.966	0.977	1.000
1.000	0.984	0.973	0.965	0.960	0.956	0.954	0.955	0.961	0.975	1.000

and the solution, after 59 iterations, is

y(0.0)	y(0.1)	y(0.2)	y(0.3)	y(0.4)	y(0.5)	y(0.6)	y(0.7)	y(0.8)	y(0.9)	y(1.0)
1.000	0.967	0.939	0.919	0.905	0.898	0.898	0.908	0.927	0.957	1.000

The iterations were stopped when the value of y at each mesh point changed from one iteration to the next by less than 10^{-5}.

Why, when the results are only given to three decimal places, was the convergence criterion ε set so small? The answer lies in the fact that if, as in this case, the rate of convergence is very low, then changes in the fourth and fifth decimal places will continue to occur for many iterations and can eventually accumulate to affect the third place. This is illustrated by the following solutions (only the portions for $0 \leq x \leq 0.5$ are shown) which were obtained using successively smaller values of ε:

ε	N	y(0.0)	y(0.1)	y(0.2)	y(0.3)	y(0.4)	y(0.5)
10^{-2}	3	1.000	0.989	0.981	0.977	0.974	0.972
10^{-3}	22	1.000	0.969	0.944	0.925	0.911	0.905
10^{-4}	41	1.000	0.967	0.940	0.919	0.905	0.898
10^{-5}	59	1.000	0.967	0.939	0.919	0.905	0.898
10^{-6}	78	1.000	0.967	0.939	0.919	0.905	0.898

N is the number of iterations to convergence. It can be seen that $\varepsilon = 10^{-5}$ is needed to obtain three-figure accuracy. Unfortunately, it is seldom possible to determine on theoretical grounds what value of ε will be required. Numerical experiments such as these are almost invariably necessary.

Similar experiments are necessary to discover whether mesh size convergence has been achieved. In other words, is $h = 0.1$ small enough? The answer to this question is 'yes – if three-figure accuracy is good enough', but the investigation is left as an exercise for the student.

What is the effect of different values of σ? The following table shows the answer:

σ	N
1.0	153
0.75	131
0.5	109
0.25	85
0	59
−0.25	30
−0.35	18
−0.4	25
−0.45	50
−0.5	(diverges)

An iteration limit of 200 was imposed. More iterations than this would have been required for $\sigma > 1.0$, but convergence would eventually have been reached. For $\sigma < -0.45$, the process diverged.

(b) To use the shooting method, the equation was written

$$y' = z \text{ with } y(0) = 1$$
$$z' = y^2 + z \text{ with } z(0) = \alpha$$

With a first guess of $\alpha = 1$, the value of $y(1)$ was found to be 4.670 47. With $\alpha = 1.01$, $y(1) = 4.701\,79$. Extrapolating to $y(1) = 1$, the new value of α was −0.171 94. Continuing, the following results were obtained:

α	$y(1)$
1	4.670 47
1.01	4.701 79
−0.171 94	1.455 22
−0.173 66	1.451 11
−0.362 42	1.009 82
−0.366 04	1.001 53
−0.366 71	1.000 00

The same solution as in part (a) was obtained.

Problems

Students are encouraged to write computer programs to implement all of the methods described here. Their programs can be tested by solving problems with known, analytical solutions – and will then be available for them to use elsewhere in their professional training and careers on problems which do *not* have known, analytical solutions. However, before they can do that, and in order to understand fully the various methods, students should work through several examples by hand; and to develop a feeling for the accuracy of the solutions being found, the examples should again have analytical solutions.

A number of problems, and their solutions, are given below. However, it is simple to generate differential equations with known solutions, and further examples can therefore easily be constructed. We merely start with the answer and work backwards to the question.

For example, to find a first-order equation of which

$$y = e^{-x} \sin x$$

is the solution, we write

$$y' = -e^{-x} \sin x + e^{-x} \cos x = -y + e^{-x} \cos x$$

The problem to be solved is therefore

$$y' + y - e^{-x} \cos x = 0 \text{ with } y(0) = 0$$

Similarly, to find a second-order equation with the same solution, we must differentiate again to obtain

$$y'' = -y' - e^{-x} \cos x - e^{-x} \sin x = -2y' - 2y$$

The second-order, initial value problem to be solved is therefore

$$y'' + 2y' + 2y = 0 \text{ with } y(0) = 0 \text{ and } y'(0) = 1$$

In each case, the initial conditions are found from the chosen solution.

This last equation can also be solved as a two-point boundary value problem:

$$y'' + 2y' + 2y = 0 \text{ with } y(0) = 0 \text{ and } y(\pi/2) = 0.20788$$

The problem has a somewhat artificial appearance – but it is nevertheless a perfectly good two-point boundary value problem. However, it is necessary to avoid generating problems with a trivial solution. For example, a change in the right-hand boundary condition to

$$y(\pi) = 0$$

results in a problem with the solution

$$y = 0$$

for all x.

It is recommended that further problems (with non-trivial solutions!) be constructed in this manner.

1. Solve the equation

$$y' = 2y + x^2 e^x \text{ with } y(0) = -2$$

for $0 < x \leq 1$.

(a) Use the simple Euler method with $h = 0.2$.
(b) Use the simple Euler method with $h = 0.1$.
(c) Use Richardson's extrapolation to improve the accuracy of the solution, recalling that the global truncation error of the Euler method is $O(h)$.
(d) Given that the analytical solution is

$$y = -(x^2 + 2x + 2)e^x$$

find the error in each case.

2. Repeat Problem 1 for the equation

$$y' + y = e^x - e^{-x} \text{ with } y(0) = \tfrac{1}{2}$$

of which the analytical solution is

$$y = \tfrac{1}{2}e^x - xe^{-x}$$

3. Solve Problems 1 and 2 by the Taylor series method.

(a) Use $h = 0.2$ and retain sufficient terms in the Taylor series to ensure that the local truncation error is less than 0.0001. How many terms are needed? What is the error in $y(1)$?
(b) Use $h = 0.5$ and the same number of terms in the series. What is now the error in $y(1)$?
(c) Use $h = 1.0$ and the same number of terms in the series. What is now the error in $y(1)$?

4. Solve

$$y' + y - e^{-x} \cos x = 0 \text{ with } y(0) = 0$$

by Milne's method for $0 < x \leq 20$.

(a) What is the largest value of the step size h which will permit the corrector formula to converge?
(b) Show (by numerical experiment) that values of h greater than this limit do not permit the solution to be obtained at all.
(c) Show that values of h smaller than the limiting value lead to solutions which exhibit an error which oscillates in sign at each step but diminishes in magnitude as $h \to 0$.

5. (a) By choosing $j = 0$ and the retention of three terms in (5.21) and four terms in (5.24), develop the Adams predictor–corrector method

$$y_{n+1} = y_n + (h/12)(23f_n - 16f_{n-1} + 5f_{n-2}) + O(h^4)$$
$$y_{n+1} = y_n + (h/24)(9f_{n+1} + 19f_n - 5f_{n-1} + f_{n-2}) + O(h^5)$$

(b) Use this method to solve

$$y' + y - e^{-x} \cos x = 0 \text{ with } y(0) = 0$$

for $0 < x \leq 20$.

(c) What is the largest value of the step size h which will permit the corrector formula to converge?

(d) Show (by numerical experiment) that values of h greater than this limit do not permit the solution to be obtained at all.

(e) Show that values of h smaller than the limiting value lead to solutions which exhibit an error which diminishes in magnitude as $h \to 0$ and which does not oscillate in sign at each step.

6. Derive estimates analogous to (5.35) for the truncation errors in the Adams and Hamming P–C methods, using the Milne predictor in each case.

7. Solve each of the following equations* and systems of equations by the various methods described, over the range $0 < x \leq 20$. Use whatever means are appropriate to estimate the local errors, and endeavour to keep them less than (i) 10^{-3}, (ii) 10^{-6} and (iii) 10^{-9}. If possible, obtain the CPU time for each solution, and compare the effort as measured by the CPU time to the respective number of function evaluations.

(a) $y' = -y^3/2$ with $y(0) = 1$ [Solution: $y = 1/(1 + x)^{1/2}$]

(b) $y' = y \cos x$ with $y(0) = 1$
(A special case of the Riccati equation.)   [Solution: $y = \exp(\sin x)$]

(c) $y' = (y/4)(1 - y/20)$ with $y(0) = 1$
(A logistic curve.)

[Solution: $y = 20/(1 + 19 \exp(-x/4))$]

(d) $y_1' = -y_1 + y_2$ with $y_1(0) = 2$
$y_2' = y_1 - 2y_2 + y_3$ with $y_2(0) = 0$
$y_3' = y_2 - y_3$ with $y_3(0) = 1$
(A linear chemical reaction.)

(e) $y_1' = 2(y_1 - y_1 y_2)$ with $y_1(0) = 1$
$y_2' = -(y_2 - y_1 y_2)$ with $y_2(0) = 3$
(The growth of two conflicting populations.)

(f) $y_1' = y_2 y_3$ with $y_1(0) = 0$
$y_2' = -y_1 y_3$ with $y_2(0) = 1$
$y_3' = -0.51 y_1 y_2$ with $y_3(0) = 1$
(Euler's equations of motion for a rigid body without external forces.)

---

* Chosen from Hull, T. E., W. H. Enright, B. M. Fellen & A. E. Sedgwick 1972. Comparing numerical methods for ordinary differential equations. *SIAM Journal of Numerical Analysis* **9**, 603–37, to which reference should be made for a valuable discussion of various methods for ordinary differential equations.

PROBLEMS

8. Solve the two-point boundary value problem

$$y'' - y + 1 = 0 \text{ with } y(0) = y(1) = 0$$

(a) Use $h_1 = 0.5$ and $h_2 = 0.25$. Apply Richardson's extrapolation to the value at the mid-point $x = 0.5$.

(b) Now repeat the calculations with successively finer meshes. Show that the extrapolated mid-point value found in (a) is not obtained by direct computation until a mesh size of $h = 0.03125$ is used.

9. Solve the two-point boundary value problem

$$y'' + e^y = 0 \text{ with } y(0) = 0, y(1) = 1$$

(a) by a finite difference approximation and (b) by the shooting method. In each case, use a step size $h = 0.1$. Note that there are *two* solutions to this problem, in which $y'(0) \approx 1.8633$ and $y'(0) \approx 9.1239$. Compare the ease (or difficulty) with which the two methods enable the two solutions to be obtained.

6

Partial differential equations I – elliptic equations

6.1 Introduction

In this chapter we start to examine some techniques used for the numerical solution of partial differential equations (PDEs) and, in particular, equations which are special cases of the linear second-order equation with two independent variables

$$a \frac{\partial^2 u}{\partial x^2} + b \frac{\partial^2 u}{\partial x \, \partial y} + c \frac{\partial^2 u}{\partial y^2} + d \frac{\partial u}{\partial x} + e \frac{\partial u}{\partial y} + fu + g = 0 \qquad (6.1)$$

where the coefficients $a, b, \ldots, g$ may be functions of x and y, but not of u.

The behaviour of the solution of (6.1) depends on the coefficients. It has been found convenient to classify the equation, and to a large extent the methods for its solution, according to the sign of the quantity $(b^2 - 4ac)$.

If $b^2 - 4ac < 0$, then (6.1) is said to be an *elliptic* differential equation. An example is

$$\nabla^2 u = \frac{\partial^2 u}{\partial x^2} + \frac{\partial^2 u}{\partial y^2} = -g \qquad (6.2)$$

which is known as *Poisson's equation*. Here, $a = c = 1$ and $b = 0$; hence $b^2 - 4ac = -4$, which is less than zero. If $g = 0$, (6.2) is called *Laplace's equation*. These equations describe, for example, the steady two-dimensional temperature distribution in a heat-conducting material, as determined by certain boundary conditions and by g (which might denote a distributed heat source in the material).

If $b^2 - 4ac = 0$, then (6.1) is said to be *parabolic*. An example is

$$a \frac{\partial^2 u}{\partial x^2} - \frac{\partial u}{\partial y} = 0 \qquad (6.3)$$

which is known as the *diffusion* or *conduction* equation. Here, $b = c = 0$ and hence $b^2 - 4ac = 0$. If x denotes a space co-ordinate and y denotes

INTRODUCTION

time, then this equation describes the one-dimensional diffusion of some quantity u, the flux of which is proportional to the local gradient $\partial u/\partial x$. For example, if u denotes temperature and a is the thermal diffusivity of some material, (6.3) describes one-dimensional heat conduction in the material as a result, say, of a change in its surface temperature. In such a case, the symbol t would normally be used to denote time, and (6.3) would be written

$$\frac{\partial u}{\partial t} = a \frac{\partial^2 u}{\partial x^2} \tag{6.3a}$$

If $b^2 - 4ac > 0$, then (6.1) is called a *hyperbolic* differential equation. An example is

$$\frac{\partial^2 u}{\partial x^2} - \frac{\partial^2 u}{\partial y^2} = 0 \tag{6.4}$$

which is known as the wave equation. Here, $b = 0$ and $a = -c = 1$, hence $b^2 - 4ac = 4$, which is greater than zero. This equation describes, for example, the displacement u of a plucked violin string, as a function of x, the distance along the string, and y, the time since the initial disturbance was created. It also has important applications in supersonic flow problems, and in other situations.

In the equations we will be considering, the coefficients a, b and c will be assumed to be constant. However, if they are variables (i.e. functions of x or y, or both), then it is possible that the class of the equation could vary throughout the solution region. If, further, they are functions of the independent variable, then the position of the boundary between regions where equations of different classification apply is not known in advance. These are complications which can occur in practice, but are beyond the present treatment.

Parabolic equations frequently arise in time-dependent problems in science and engineering. For example, the motion of a viscous fluid is governed by the *Navier–Stokes equations*, one of which, under appropriate conditions, is

$$\frac{\partial u}{\partial t} + u\frac{\partial u}{\partial x} + v\frac{\partial u}{\partial y} + w\frac{\partial u}{\partial z} = -\frac{1}{\rho}\frac{\partial p}{\partial x} + \nu\left(\frac{\partial^2 u}{\partial x^2} + \frac{\partial^2 u}{\partial y^2} + \frac{\partial^2 u}{\partial z^2}\right) \tag{6.5}$$

where u, v and w are the velocity components of the fluid in the x-, y- and z-directions, respectively; p, ρ and ν are the pressure, density and kinematic viscosity, respectively, of the fluid; and t denotes time.

If this equation (and other necessary equations) are used to describe the unsteady flow pattern as a fluid from an emptying reservoir moves along a pipe, in the x-direction, then the equation can be regarded as being parabolic in this direction, which is the predominant direction of flow (velocities *across* the pipe will be small compared with those *along* the pipe). This can be verified by comparing (6.1) with (6.3), in which y is replaced by t.

On the other hand, if the flow is steady, so that $\partial u/\partial t = 0$, the flow is elliptic; it is in essence similar to a three-dimensional version of (6.2).

Equation (6.5) is *non-linear*, because of the appearance of the velocity components as coefficients: the second term, for instance, can be written $\frac{1}{2}\partial u^2/\partial x$. It is only one of four equations which are needed to find the four unknowns (u, v, w and p) in this problem. Also, these four equations are *coupled*, because all of the unknowns appear in three of the equations, and three of them appear in the fourth equation*.

Methods for the solution of these equations are beyond the scope of this book. Indeed, they are the subject of intense research activity. However, we shall look at some simple methods for simplified versions of (6.5). These methods underlie the advanced techniques for a study of the full equations; and the simplified equations in their own right have applications in science and engineering.

In this chapter we shall consider some introductory methods for the solution of elliptic problems, and in the next chapter we shall look at parabolic problems. These types of problem can be solved by finite difference methods. Generally, other methods – notably, the method of characteristics – are used for hyperbolic equations such as (6.4), and consideration of them will not be given in this book. Finite difference methods such as those described here can sometimes also be used for hyperbolic equations. A discussion of this possibility is confined to a worked example in the next chapter.

6.2 The approximation of elliptic equations

Suppose that a thin rectangular sheet of metal, of dimensions X and Y, is subjected to some specified thermal condition along its edges, and suppose further that (perhaps because the metal is carrying an electric current) heat is generated in the sheet at a rate proportional to g, which we will allow to be a function of x or y, or both. Then (6.2) gives the distribution of temperature, T, as a function of x and y.

We will seek a solution of the problem by constructing a finite difference approximation (FDA) to (6.2). As shown in Figure 6.1, we superimpose a rectangular mesh on the solution region, and will try to find the values of T at the *nodes* or *mesh points*. There are $M = X/h_x$ intervals in the x-direction and $N = Y/h_y$ intervals in the y-direction, where h_x and h_y are the respective

---

* The fourth equation is the *continuity equation*. Students of fluid dynamics will know this; and for others it is not important. Incidentally, if the fluid is *compressible*, the density ρ is also variable, and a fifth equation – an equation of state – will be needed. If, in addition, the temperature of the fluid is a variable, then the *energy* equation is required.

THE APPROXIMATION OF ELLIPTIC EQUATIONS 213

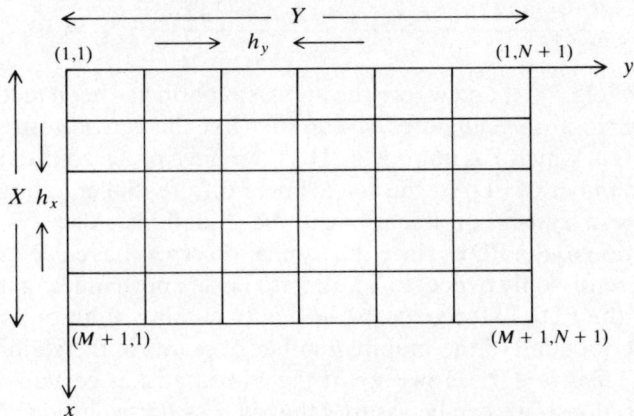

Figure 6.1 Mesh points for the solution of Poisson's equation.

mesh sizes. A double subscript notation is used to give the co-ordinates of any point in the mesh. Thus,

$$T_{i,j} \equiv T[(i-1)h_x, (j-1)h_y] \tag{6.6}$$

where i and j are the row and column numbers of the mesh point, as shown in the figure. Since there are M and N intervals in the two directions, i and j range from 1 to $(M+1)$ and from 1 to $(N+1)$, respectively. If the axes are located in the somewhat unconventional manner shown in the figure, then the row and column numbers of an element in the $(M+1) \times (N+1)$ matrix of values of T bear a direct relationship to the x- and y-co-ordinates of the corresponding point in space, which makes the conceptual connection between the subscripts and the physical location of the point somewhat easier to understand.

Using the notation of (6.6), the simplest central difference approximation to $\partial^2 T/\partial x^2$ at a point (i, j) is

$$\frac{\partial^2 T}{\partial x^2} = \frac{T_{i+1,j} - 2T_{i,j} + T_{i-1,j}}{h_x^2} + O(h_x^2) \tag{6.7}$$

Similarly,

$$\frac{\partial^2 T}{\partial y^2} = \frac{T_{i,j+1} - 2T_{i,j} + T_{i,j-1}}{h_y^2} + O(h_y^2) \tag{6.8}$$

To simplify the presentation, it will now be assumed that $h_x = h_y = h$. It is emphasized that this is *not* necessary, and is often not true (although in practice it is found undesirable, from considerations of both accuracy and stability, to have h_x and h_y very different from each other). Then the insertion of (6.7) and (6.8) into (6.2) yields the FDA

$$\frac{T_{i+1,j} + T_{i-1,j} + T_{i,j+1} + T_{i,j-1} - 4T_{i,j}}{h^2} + g_{i,j} = 0 \qquad (6.9)$$

where $g_{i,j} \equiv g(x_i, y_j)$, and where the approximation has been made that the truncation errors are negligible. An equation like this can be written for each of the points at which T is unknown. Thus, we have replaced the single PDE (6.2) by a number of FDAs: the *single* linear differential equation has been replaced by a *system* of linear algebraic equations. The system is an approximation to the PDE, since the truncation errors have been neglected. We can therefore only expect its solution to be an approximation to the true solution of the PDE. Moreover, we will only obtain a solution at the mesh points, and not at all of the infinite number of points in the solution region $0 \leq x \leq X$, $0 \leq y \leq Y$. However, if the FDA satisfies certain conditions [which (6.9) does], we can be assured that as $h \to 0$ its solution approaches that of the PDE. Thus, we can make the numerical solution as accurate as we like by making h sufficiently small, the only limits being computer time and memory.

6.3 Boundary conditions

In order that a solution of (6.2) can be found, boundary conditions for T must be given. These must be translated into conditions for $T_{i,j}$ in order that a solution can be found for the system represented by (6.9).

Three types of boundary condition are commonly encountered:

(a) $T = f(x, y)$;
(b) $\partial T/\partial n = f(x, y)$, where n denotes the co-ordinate direction normal to the boundary;
(c) $pT + q\, \partial T/\partial n = f(x, y)$, i.e. a combination of the first and second types.

Boundary conditions of type (a) are known as Dirichlet conditions, those of type (b) as Neumann conditions and those of type (c) as mixed conditions. Between them they cover most of the boundary conditions encountered in practice (and the only conditions which may apply to the second-degree equations we shall be considering).

In the heat conduction problem of the previous section these three conditions correspond, respectively, to a specified temperature, a specified heat flux, and a specified heat transfer coefficient at the edges of the plate.

If the first boundary condition applies, then all boundary values of T are known; viz. $T_{1,1}, T_{1,2}, \ldots, T_{1,N+1}, T_{2,1}, T_{3,1}, \ldots, T_{M+1,1}, T_{M+1,2}, T_{M+1,3}, \ldots, T_{M+1,N+1}, T_{2,N+1}, T_{3,N+1}, \ldots, T_{M,N+1}$. Thus, it is only at the $(M-1)(N-1)$ *internal* mesh points that T is not known, and (6.9) then represents a system of $(M-1)(N-1)$ equations in the same number of unknowns. We will consider this type of boundary condition first.

6.4 Non-dimensional equations again

Suppose we require the solution of

$$\frac{\partial^2 T}{\partial x^2} + \frac{\partial^2 T}{\partial y^2} + g(x, y) = 0 \tag{6.10}$$

in the region $0 \leq x \leq X, 0 \leq y \leq Y$, subject to the conditions

$$\begin{aligned} T(0, y) &= A & \text{for } 0 \leq y \leq Y \\ T(X, y) &= B & \text{for } 0 \leq y \leq Y \\ T(x, 0) &= C & \text{for } 0 \leq x \leq X \\ T(x, Y) &= D & \text{for } 0 \leq x \leq X \end{aligned} \tag{6.10a}$$

where, for simplicity, A, B, C and D are constants, and where g is a given function of x and y.

There are six parameters involved in this problem – X, Y, A, B, C and D – in addition to the function g (which will vary from problem to problem), and they must all be specified before the solution can be found. Moreover, if one (or more) is changed, then the solution process must be repeated. We can generalize the solution, and reduce the number of parameters, by transforming (6.10) into a *non-dimensional equation*.

Suppose, for the sake of discussion, that A and C are, respectively, the smallest and largest of the four boundary values of T. Then if $g = 0$, it can be shown that $A \leq T \leq C$ everywhere in the solution region. Thus, if a new scaled variable θ is defined by

$$\theta = \frac{T - A}{C - A} \tag{6.11}$$

then θ will be non-dimensional (being a ratio of two temperature differences in this case). Moreover, when $T = A$, then $\theta = 0$, and when $T = C$, then $\theta = 1$. Thus $0 \leq \theta \leq 1$ in the solution region. It is convenient to have θ of the order of unity, since this makes spurious values of θ more readily detectable, and where possible the method of scaling should be chosen with this objective in mind. A further advantage is that the possibility of computer overflow or underflow is reduced.

If g is *not* zero everywhere, then θ need not lie between 0 and 1, but (6.11) is still a convenient substitution.

We also define dimensionless co-ordinates, in terms of either X or Y (it usually makes no difference which is chosen). Thus, let

$$\xi = x/X \qquad \eta = y/X$$

for example. Then

$$\frac{\partial T}{\partial x} = \frac{\partial}{\partial \xi} \{\theta(C - A) + A\} \frac{d\xi}{dx} = \frac{C - A}{X} \frac{\partial \theta}{\partial \xi}$$

(ξ is not a function of y, therefore the ordinary derivative $d\xi/dx$ is appropriate) and

Similarly,

$$\frac{\partial^2 T}{\partial x^2} = \frac{\partial}{\partial x}\left(\frac{\partial T}{\partial x}\right) = \frac{\partial}{\partial \xi}\left(\frac{C-A}{X}\frac{\partial T}{\partial \xi}\right)\frac{d\xi}{dx} = \frac{C-A}{X^2}\frac{\partial^2 \theta}{\partial \xi^2}$$

$$\frac{\partial T}{\partial y} = \frac{\partial}{\partial \eta}\{\theta(C-A) + A\}\frac{d\eta}{dy} = \frac{C-A}{X}\frac{\partial \theta}{\partial \eta}$$

and

$$\frac{\partial^2 T}{\partial y^2} = \frac{\partial}{\partial y}\left(\frac{\partial T}{\partial y}\right) = \frac{\partial}{\partial \eta}\left(\frac{C-A}{X}\frac{\partial \theta}{\partial \eta}\right)\frac{d\eta}{dy} = \frac{C-A}{X^2}\frac{\partial^2 \theta}{\partial \eta^2}$$

The function $g(x, y)$ will transform into another function, say $G(\xi, \eta)$. For example, suppose

$$g(x, y) = x + y^2$$

then

$$g(x, y) = \xi X + \eta^2 X^2 = G(\xi, \eta)$$

Thus (6.10) becomes

$$\frac{C-A}{X^2}\frac{\partial^2 \theta}{\partial \xi^2} + \frac{C-A}{X^2}\frac{\partial^2 \theta}{\partial \eta^2} + G(\xi, \eta) = 0$$

or

$$\frac{\partial^2 \theta}{\partial \xi^2} + \frac{\partial^2 \theta}{\partial \eta^2} + \Gamma(\xi, \eta) = 0 \qquad (6.12)$$

say, in $0 \leq \xi \leq 1$, $0 \leq \eta \leq R$, where $R = Y/X$ is the *aspect ratio* of the solution region, and Γ is now a function of A, C and X. The boundary conditions become

$$\begin{aligned}
\theta(0, \eta) &= 0 &&\text{for } 0 \leq \eta \leq R \\
\theta(1, \eta) &= E &&\text{for } 0 \leq \eta \leq R \\
\theta(\xi, 0) &= 1 &&\text{for } 0 \leq \xi \leq 1 \\
\theta(\xi, R) &= F &&\text{for } 0 \leq \xi \leq 1
\end{aligned} \qquad (6.12a)$$

where $E = (B - A)/(C - A)$ and $F = (D - A)/(C - A)$.

In the problem represented by (6.12), there are now only three parameters – the aspect ratio R, and the boundary conditions E and F – compared with the previous six, together with the function Γ, and the problem has been greatly simplified and generalized.

The detailed manner in which non-dimensional variables are formed will depend on the particular problem, and this approach may not be applicable to other situations. The construction of a non-dimensional equation should always be undertaken before a numerical solution is sought.

6.5 Method of solution

We now turn to the solution of (6.12) and (6.12a). We will use a mesh size of h (the same in each direction), so that there are $M = 1/h$ intervals in the x-direction and $N = R/h$ intervals in the y-direction. The notation and boundary conditions are shown in Figure 6.2. It is only at the *internal* mesh points, enclosed by the broken line, that θ is unknown.

It will be realized that the size of the system represented by (6.9) can be very large. It is generally found that, to achieve an acceptable accuracy, h should be not more than 0.1, and often needs to be very much less. Thus, for $R = 1$ we might have an 11×11 mesh; there are therefore $9 \times 9 = 81$ *internal* mesh points, and (6.9) represents a system of 81 equations in 81 unknowns. If, for greater accuracy, we were to use $h = 0.01$, then the size of the system would increase to $99 \times 99 = 9801$.

We started with a linear PDE, and now have a large linear system of algebraic equations. As discussed in Chapter 3, a good method for solving such a system is Gauss–Seidel iteration or, if we can find ω_{opt}, successive over-relaxation (SOR). Thus (6.9) is written

$$\theta_{i,j}^{k+1} = \theta_{i,j}^{k} + (\omega/4)(\theta_{i+1,j}^{k} + \theta_{i-1,j}^{k+1} + \theta_{i,j+1}^{k} + \theta_{i,j-1}^{k+1} - 4\theta_{i,j}^{k} + h^2 \Gamma_{i,j}) \quad (6.13)$$

for $2 \leq i \leq M$, $2 \leq j \leq N$; k is the iteration counter, and it has been assumed that the iterations are performed in order of increasing row and column number. Thus the $(k + 1)$th estimates of the values of $\theta_{i-1,j}$ and $\theta_{i,j-1}$ will already have been computed before the $(k + 1)$th estimate of $\theta_{i,j}$ is found.

In order to start, an initial guess, $\theta_{i,j}^0$ is required. Although the number of iterations to convergence depends on the quality of the initial guess, it is not normally worth going to a great deal of trouble to make a good one. Typically, we might choose $\theta_{i,j}^0 = 0$, or 1, or perhaps $(1 + E + F)/4$, i.e. the average of the four boundary values. This initial guess only applies to the *internal* mesh points; the boundary values are set at the beginning of the

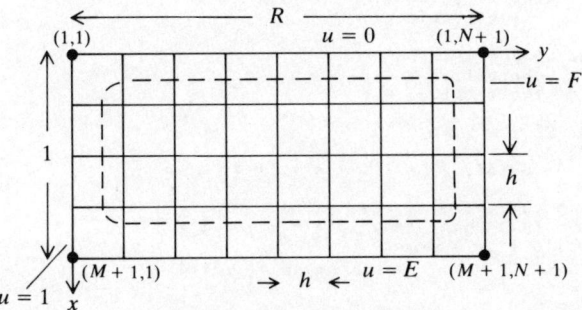

Figure 6.2 Problem notation.

calculations and are not altered. They are automatically taken into account when (6.13) is applied for $i = 2$ or M, and for $j = 2$ or N.

For Poisson's (or Laplace's) equation in a rectangular region, it can be shown that the optimum over-relaxation factor is given by the smaller root of the quadratic equation

$$t^2\omega^2 - 16\omega + 16 = 0$$

where

$$t = \cos(\pi/m) + \cos(\pi/N)$$

In such cases, SOR is a simple and economical method.

Figure 6.3 shows the Fortran listing of a program for the solution of this problem, and the result of executing the program. In this example, the aspect ratio $r = 2$ and $\Gamma(\xi, \eta) = 5$. The boundary condition values are $E = F = 0.5$.

Figure 6.3 The solution of $\nabla^2\theta + 5 = 0$ by SOR.

```
c
            dimension  theta(21,41)
c
            r = 2.
            e = 0.5
            f = e
            gamma = 5.
            pi = 3.14159
c
      5     write (*,10)
     10     format (/' how many intervals along the xi-axis ?   ',$)
            read (*,*) m
            if (m .le. 0) stop
            h = 1./m
            n = r * m
            mp1 = m + 1
            np1 = n + 1
            mmid = 1 + m/2
            nmid = 1 + n/2
c
c           set boundary conditions and initial guess
c
            do 15 i = 1 , mp1
               theta(i,1) = 1.0
     15        theta(i,np1) = f
c
            do 20 j = 1 , np1
               theta(1,j) = 0.0
     20        theta(mp1,j) = e
c
            do 25 i = 2, m
            do 25 j = 2, n
     25        theta(i,j) = (1. + e + f)/4.
c
            t = cos (pi/m) + cos (pi/n)
            w = (8. - sqrt (64.-16.*t*t) ) / t / t
            write (*,27) w
     27     format (/' sor factor omega = ',f7.4)
            w4 = w / 4
            k = 0
c
c           iteration loop starts here
```

METHOD OF SOLUTION

Figure 6.3 (*continued*)

```
c
   30  bit = 0.0
       do 35 i = 2, m
       do 35 j = 2, n
          thetach = w4 * ( theta(i+1,j) + theta(i-1,j) + theta(i,j+1)
      1              + theta(i,j-1) - 4. * theta(i,j) + gamma*h*h )
          bit = bit + abs(thetach)
   35     theta(i,j) = theta(i,j) + thetach
       bit = bit / (m-1)/(n-1)
       if (bit .ge. 0.0001) then
c
c      solution has not converged
c
          k = k + 1
          if (k .lt. 100) then
              go to 30
          else
c
              write (*,65)
   65         format(/' *** failed to converge in 100 iterations *** ')
              go to 43
          endif
       endif
c
c      solution has converged
c
       write (*,40) k
   40  format (/' converged in ',i4,' iterations '/)
   43  if (h .lt. 0.2) go to 55
c
       do 45 i = 1, mp1
   45      write (*,50) (theta(i,j), j = 1, np1)
   50      format (9f6.3)
c
   55  write (*,60) theta (mmid, nmid)
   60  format (/' mid-point value is ',f8.4)
       go to 5
       end
```

Typical output from this program appears as follows:

```
how many intervals along the xi-axis ?   2

sor factor omega =   1.0334

converged in    4 iterations

 .000   .000   .000   .000   .000
1.000   .902   .857   .777   .500
 .500   .500   .500   .500   .500

mid-point value is .    .8571

how many intervals along the xi-axis ?   4

sor factor omega =   1.2668

converged in    9 iterations

 .000   .000   .000   .000   .000   .000   .000   .000   .000
1.000   .732   .645   .611   .592   .575   .551   .518   .500
1.000   .971   .925   .894   .871   .845   .800   .708   .500
1.000   .915   .877   .856   .840   .820   .783   .701   .500
 .500   .500   .500   .500   .500   .500   .500   .500   .500

mid-point value is     .8707

how many intervals along the xi-axis ?   8
```

(*continued*)

```
sor factor omega =   1.5325
converged in    18 iterations

mid-point value is    .8734
how many intervals along the xi-axis ?   16
sor factor omega =   1.7323
converged in    33 iterations

mid-point value is    .8738
how many intervals along the xi-axis ?   0
```

Results are shown in full for $h = 0.5$ and $h = 0.25$; for $h = 0.125$ and $h = 0.0625$, the table of values of θ has been omitted to save space, and only the mid-point value is given.

In Section 3.8 we discussed the *convergence* of the iterative solution of a system of equations, i.e. how to decide when the solution has been obtained to within some acceptable degree of accuracy. To check on convergence in this example, the sum of the absolute values of the changes in θ at each internal point is found. This quantity is called 'bit' in the program. The average value for the $(M - 1)(N - 1)$ internal points is then compared with a small number, here chosen to be 0.0001. Often a *relative* convergence test is used, but it is clear here that the solution is going to be of the order of unity everywhere, and the simpler test is therefore adequate.

The program stops if the number of iterations exceeds 100. A limit like this should always be included, even when problems are not really expected partly because during development of the program, you might make a mistake which prevents convergence of the iterations.

It will be noted that a singularity can exist at a corner of the rectangular solution region, where the boundary conditions can conflict. For example (see Fig. 6.2), if $\theta = 0$ at $x = 0$ and $\theta = 1$ at $y = 0$, what is the value of θ at the origin? This is 'only' a mathematical singularity – in practice, a discontinuity like this, which requires an infinite temperature gradient, cannot exist. Nevertheless, it is possible for very rapid changes in θ to occur near the corner, and it is valid to ask what value of θ should be used there. In fact, using the differencing scheme described, it does not matter what value is chosen; the corner points are not used in the iterative procedure. As shown in Figure 6.4, the calculation of $\theta(2, 2)$ requires values for $\theta(1, 2)$ and $\theta(2, 1)$, but no calculation requires $\theta(1, 1)$. We are therefore able to sidestep this apparent difficulty.

In the neighbourhood of such a discontinuity we would nevertheless expect to lose accuracy, especially if a coarse mesh is used. It can be seen that by refining the mesh from 1/8 to 1/16 a change in the mid-point value from

THE ACCURACY OF THE SOLUTION

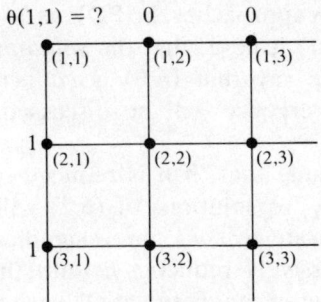

Figure 6.4 Boundary conditions near a corner.

0.8734 to 0.8738, or about 0.05%, results. The error in the mid-point value will also be of this order of magnitude. This suggests that the mesh is now probably fine enough. A smaller change can be expected following further mesh refinement and the additional accuracy will be small compared with the cost of the additional computations.

6.6 The accuracy of the solution

It is important to be aware of the distinction between the solution of the FDA and the solution of the PDE, and to realize what is meant by 'the accuracy of the solution'.

The accuracy of the solution of the FDA, i.e. of (6.9), can be improved by requiring bit to be as small as we like – at any rate down to the limit of accuracy of the computer being used. We say that the iterative procedure defined by (6.13) has *converged* when 'bit' is less than (in the present example) 0.0001. If this value were reduced to 10^{-10} and double-precision arithmetic used, then we could eventually obtain what is essentially an *exact* solution of (6.9). However, this would not be an exact solution of (6.2), because (6.9) is only an approximation to (6.2). In deriving (6.9) we have neglected the truncation errors of the finite difference approximations, which are of the order h^2. In order that the solution of (6.9) should approach the solution of (6.2), it is necessary that h should become small, i.e. that the number of mesh points should become large.

We cannot easily obtain a useful estimate of the magnitude of the truncation errors, or of the differences between the solutions of (6.9) and (6.2). We know that these errors are approximately proportional to h^2, but we cannot obtain the factor of proportionality. So how can we know that we have a 'good' solution of (6.2)?

Since we know that the truncation error of (6.9) is of order h^2, we also know that the error reduces as h reduces. We say that (6.9) is *consistent* with

(6.2), i.e. that the FDA approaches the PDE as $h \to 0$. It can also be shown, although we will not do it here, that the *solution* of (6.9) approaches the *solution* of (6.2)*: we say that (6.9) is *convergent* to (6.2) as $h \to 0$. Consistency and convergence will be discussed more fully in the next chapter.

Thus, we can conclude that, if h is reduced, then the accuracy of the computed values of $\theta_{i,j}$ as solutions of (6.2) will improve. If we want a solution of (6.2) accurate to, say, three significant figures, then we can achieve this by successively reducing h until the solution of (6.9) only changes in the fourth significant figure at all mesh points.

Here we see again the double meaning of the word 'convergence': the approach of successive estimates of θ to the final solution in an iterative procedure (achieved by making 'bit' sufficiently small) and also the approach of the solution of the FDA to the solution of the PDE (which we achieve by reducing h).

A reduction of h implies an increase in the number of mesh points, and therefore in the amount of computer storage required. It also implies an increase in the amount of arithmetic involved in the solution procedure. There is therefore a practical limit on h, dependent on the capacity of the computer and on the amount of time (i.e. money) we are prepared to spend on the solution.

6.7 Use of Richardson's extrapolation

In Chapter 4 we saw how a knowledge of the form of the truncation error could be used to improve the accuracy of a numerical integration, and in Chapter 5 the same idea was applied to the finite difference solution of ordinary differential equations. The procedure known as Richardson's extrapolation is also useful in the solution of partial differential equations.

Two solutions must be found: the first uses a mesh size of h_1, and the second uses a mesh of, typically, half the size, viz. $h_2 = h_1/2$. Every point in the first (coarser) mesh will also be a point in the second mesh, and at these points two estimates $\theta^1_{i,j}$ and $\theta^2_{i,j}$ are obtained. Since we know that the error is proportional to h^2, (4.59) tells us that an improved estimate of $\theta_{i,j}$ is given by

$$\theta_{i,j} = \theta^2_{i,j} + (\theta^2_{i,j} - \theta^1_{i,j})/3$$

---

* The fact that (6.9) approaches (6.2) does not necessarily mean that the *solution* of (6.9) approaches the *solution* of (6.2) as $h \to 0$. However, this *is* the case with a well-constructed FDA, and will be assumed to be true in all of the examples considered here.

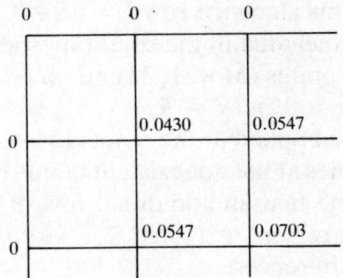

Figure 6.5 The solution of $\nabla^2 u + 1 = 0$ with $h = 0.25$.

Worked example

We can see the power of this procedure by solving the problem

$$\nabla^2 \theta + 1 = 0$$

in the unit square $0 \leq x, y \leq 1$, with $\theta = 0$ on all four boundaries.

Using $h = 0.5$, there is only one internal mesh point, at $(0.5, 0.5)$, where we find $\theta^1 = 0.0625$. With $h = 0.25$, and taking advantage of the symmetry about the lines $x = 0.5$ and $y = 0.5$, we obtain the values shown in Figure 6.5, and in particular the centre point value $\theta^2 = 0.0703$. The 'extrapolated' value is thus

$$0.0703 + (0.0703 - 0.0625)/3 = 0.0729$$

The analytical solution at the mid-point can be found: it is 0.0736. The extrapolated value compares very favourably with this.

We also notice from the example that, using $h = 0.5$, the error at the mid-point is 0.0111, while with $h = 0.25$ it is 0.0033. As h is reduced by a factor of two, the error is reduced by a factor of 3.4. Theory predicts that the error reduction should be $2^2 = 4$, but we can only expect that to apply for small values of h, and $h = 0.5$ is, in fact, the largest possible value. The agreement (between 3.4 and 4) is, under these circumstances, remarkably good!

6.8 Other boundary conditions

We shall now modify the boundary conditions of the problem so that a Neumann condition of type (b) of Section 6.3 applies on the edge $x = 1$. Suppose that this edge is being heated or cooled by some external device, so that at $x = 1$

$$\partial \theta / \partial x = S = \text{constant} \tag{6.14}$$

The values of θ at points along the row $I = M + 1$ are now not known, and these points must be included in the solution procedure. We will assume* that values of θ at the points $(M + 1, 1)$ and $(M + 1, N + 1)$ are still given by the conditions on $y = 0$ and $y = R$.

Thus (6.13) must be applied at the points $(M + 1, J)$, where $J = 2, \ldots, N$, which requires values at the nonexistent points $(M + 2, J)$. As shown in Figure 6.6, we imagine that an additional row of fictitious mesh points is added to the solution region, at $I = M + 2$. We now replace (6.14) by its FDA, using central differences:

$$\frac{\theta_{M+2,j} - \theta_{M,j}}{2h} = S + O(h^2) \qquad (6.15)$$

or

$$\theta_{M+2,j} = \theta_{M,j} + 2hS \qquad (6.16)$$

neglecting the truncation error. Then, along the row $I = M + 1$ the use of (6.16) in (6.13) yields

$$\theta_{M+1,j}^{k+1} = \theta_{M+1,j}^{k} + (\omega/4)(2\theta_{M,j}^{k+1} + 2hS + \theta_{M+1,j-1}^{k+1} + \theta_{M+1,j+1}^{k} - 4\theta_{M+1,j}^{k}) + h^2 g_{M+1,j} \qquad (6.17)$$

for $2 \leq j \leq N$.

An extra step must be added to the iteration procedure, corresponding to (6.17).

Notice that (6.15) is a second-order approximation, the same order as the main iteration step (6.13). A slightly simpler procedure can be derived by replacing (6.14) by the backward formula

$$\frac{\theta_{M+1,j} - \theta_{M,j}}{h} = S + O(h)$$

or

$$\theta_{M+1,j} = \theta_{M,j} + hS \qquad (6.18)$$

---

* Here we have a conflict of boundary conditions at the corners, and must make a reasonable assumption based on our knowledge of the physical problem.

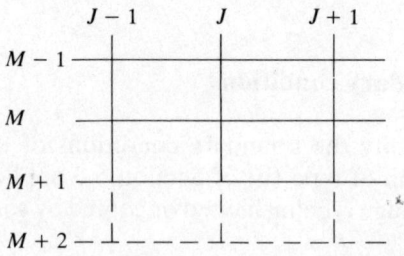

Figure 6.6 Fictitious mesh points for a derivative boundary condition.

After each iteration stage has been completed, values of θ along the bottom row could be computed using (6.18) for $2 \leq j \leq N$, but the errors in the boundary values found in this manner are greater than those of the internal points. Therefore this procedure, although simpler, is not recommended.

Type (c) boundary conditions of Section 6.3 are treated in the same way. If (6.14) is replaced by

$$p\theta + q\frac{\partial \theta}{\partial x} = S$$

at $x = 1$, then corresponding to (6.15) we have

$$p\theta_{M+1,j} + \frac{q}{2h}(\theta_{M+2,j} - \theta_{M,j}) = S + O(h^2)$$

or

$$\theta_{M+2,j} = \theta_{M,j} + \frac{2h}{q}(S - p\theta_{M+1,j}) \qquad (6.19)$$

Then, along the row $i = M + 1$ (6.13) becomes

$$\theta_{M+1,j}^{k+1} = \theta_{M+1,j}^{k} + \frac{\omega}{4}\left\{2\theta_{M,j}^{k+1} + \theta_{M+1,j-1}^{k+1} + \theta_{M+1,j+1}^{k} - 4\theta_{M+1,j}^{k}\right.$$
$$\left. - \left(4 + \frac{2hp}{q}\right)\theta_{M+1,j}^{k} + \frac{2hS}{q} + h^2 g_{M+1,j}\right\} \qquad (6.20)$$

An extra step must be added to the iteration procedure corresponding to (6.20). Again, this is a second-order approximation and is the recommended process.

6.9 Relaxation by hand-calculation

It may sometimes be desirable, and adequate, to find an approximate solution of an elliptic equation by a manual iterative calculation. If a coarse mesh is used, so that there are no more than about a dozen mesh points, and if care is taken in the selection of the initial estimate of the solution, then the work will not be excessive.

A finite difference approximation of Poisson's equation

$$\frac{\partial^2 \theta}{\partial x^2} + \frac{\partial^2 \theta}{\partial y^2} + G = 0 \qquad (6.21)$$

in which, for simplicity, G will be assumed constant, is

$$\theta_{i+1,j} + \theta_{i-1,j} + \theta_{i,j+1} + \theta_{i,j-1} - 4\theta_{i,j} + h^2 G = 0 \qquad (6.22)$$

where, again for simplicity, $h_x = h_y = h$ has been used.

If the values of $\theta_{i,j}$ do *not* satisfy (6.22), then the quantity $R_{i,j}$ defined by

$$R_{i,j} = \theta_{i+1,j} + \theta_{i-1,j} + \theta_{i,j+1} + \theta_{i,j-1} - 4\theta_{i,j} + h^2 G \qquad (6.23)$$

will not be zero. $R_{i,j}$ is known as the *residual* at the point (i, j), and it is desired to make $|R_{i,j}|$ as small as possible for all values of i and j.

Suppose that the value of $\theta_{i,j}$ is altered by $\delta\theta$. Then (6.23) shows that $R_{i,j}$ will be altered by $-4\delta\theta$. Moreover, $\theta_{i,j}$ will appear in the equations for $R_{i+1,j}$, $R_{i-1,j}$, $R_{i,j+1}$ and $R_{i,j-1}$. For example,

$$R_{i+1,j} = \theta_{i+2,j} + \theta_{i,j} + \theta_{i+1,j+1} + \theta_{i+1,j-1} - 4\theta_{i+1,j} + h^2 G$$

Therefore each of these quantities will be altered by $\delta\theta$ when $\theta_{i,j}$ is altered by $\delta\theta$. In other words, if the value of θ at any mesh point is, say, *increased* by $\delta\theta$, then the residual at that point is *decreased* by $4\delta\theta$ and the residuals at the four surrounding mesh points are each *increased* by $\delta\theta$. This knowledge may be used to modify the distribution of θ selectively to minimize the values of R.

The process is similar to the method of SOR discussed above, except that it is now proposed to do the calculations manually (i.e. using a hand-held calculator rather than a computer) so that we can exercise some discretion and judgement in the course of the calculations.

At the boundaries where θ is specified, relaxation is not to be applied. At boundaries where the gradient of θ is specified, (6.23) must be modified.

Suppose that $\partial\theta/\partial x = H$ at the point $(i, 1)$ of boundary AB as shown in Figure 6.7. Introducing the fictitious mesh point $(i, 0)$, the boundary condition yields

$$\theta_{i,0} = \theta_{i,2} - 2hH$$

Hence,

$$\begin{aligned} R_{i,1} &= \theta_{i,2} + \theta_{i,0} + \theta_{i+1,1} + \theta_{i-1,1} - 4\theta_{i,1} + h^2 G \\ &= 2\theta_{i,2} + \theta_{i+1,1} + \theta_{i-1,1} - 4\theta_{i,1} + h^2 G - 2hH \end{aligned} \qquad (6.24)$$

Thus a change of $\delta\theta$ at the point $(i, 2)$ will cause a change of $2\delta\theta$ at the boundary point $(i, 1)$, and a change of $\delta\theta$ at the point $(i, 1)$ will, as before, cause a change of $-4\delta\theta$ at the point $(i, 1)$.

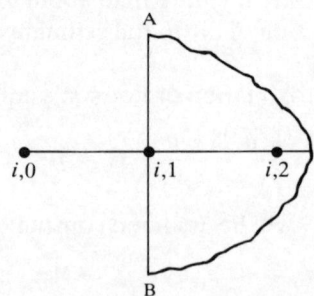

Figure 6.7 A fictitious mesh point near a boundary.

Points which are on a line of symmetry also require special treatment. Suppose that $(i, j - 1)$ and $(i, j + 1)$ are points on either side of a line of symmetry passing through the point i, j. Since a change in θ of $\delta\theta$ at $(i, j - 1)$ must be accompanied by an identical change at $(i, j + 1)$ to retain symmetry, the combined effect of these changes on $R_{i,j}$ will be $2\delta\theta$. This result also follows from the foregoing treatment of boundary points. A line of symmetry is equivalent to a boundary at which the normal derivative H is zero.

Worked example

As an example, let us solve*

$$\frac{\partial^2 T}{\partial x^2} + \frac{\partial^2 T}{\partial y^2} + 100 = 0$$

in the rectangle $0 \leq x \leq 3$, $0 \leq y \leq 6$, subject to the conditions $T = 0$ at $x = 0$, $T = 200$ at $x = 3$, $\partial T/\partial y = 100$ at $y = 0$ and $\partial T/\partial y = -100$ at $y = 6$.

Since the equation and boundary conditions are symmetrical about $y = 3$, we will be able to consider just half of the solution region (say $0 \leq y \leq 3$), with the new condition $\partial T/\partial y = 0$ at $y = 3$. We will use a grid size $h = 1$.

The method of solution is as follows. It requires a large sheet of paper, a soft pencil and an eraser.

(1) The solution region is drawn, showing the grid lines and points at a spacing of at least 2 cm.
(2) The given boundary values are written on the drawing adjacent to the relevant grid points.
(3) Estimates of the solution at all other grid points are then made and entered on the drawing.
(4) The residuals at the grid points at which T is unknown are calculated and entered.

The solution as it appears at this stage is shown in Figure 6.8 (which has been drawn with the axes in their conventional orientation). In this diagram the estimate of T at each grid point has been written just above and to the left of the point, while the corresponding residual is above and to the right. The initial estimate has been made that $T = 200$ at $x = 1$ for all y, and that $T = 250$ at $x = 2$ for all y. The residuals are calculated using (6.23) at the points labelled 'A' and using (6.24) at the points labelled 'B'. At boundary points labelled 'C', T is given and residuals are not calculated. Notice that the initial values of several residuals are zero.

---

* In this example dimensional quantities, rather than non-dimensional quantities, are used. This avoids the necessity of using decimal fractions, and makes the presentation somewhat tidier.

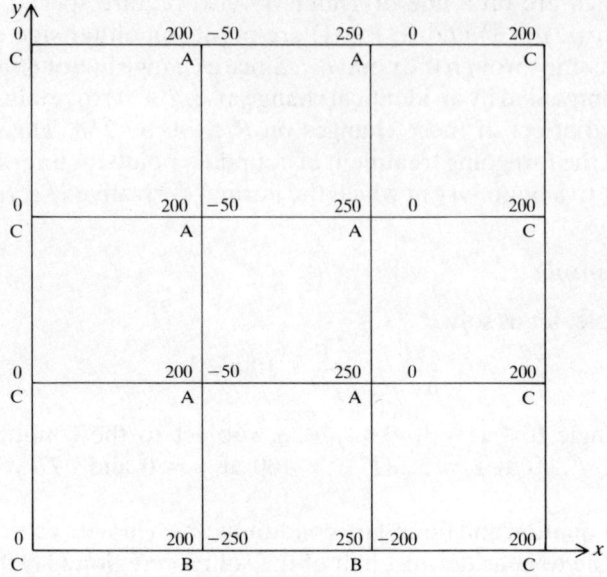

Figure 6.8 Relaxation by hand-calculation: initial conditions.

(5) The point at which the residual is the largest (in absolute value) is now selected. In this case, it is the point (2, 1), where $R = -250$.

(6) An attempt is now made to eliminate (i.e. to make zero) the residual at (2, 1) by altering $T_{2,1}$. Since a change of δT at (2, 1) will alter $R_{2,1}$ by $-4\delta T$, we choose to *reduce $T_{2,1}$ by 65*, thereby *increasing $R_{2,1}$ by $4 \times 65 = 260$* from -250 to $+10$. Simultaneously, the residuals at (2, 2) and (3, 1) are reduced by 65. Thus, the result of reducing $T_{2,1}$ by 65 is to alter the residuals at two of the neighbouring points to $\dot{R}_{2,2} = -115$ and $R_{3,1} = -265$. These changes are entered on the diagram by recording the *change* in $T_{2,2}$ and the *new values* of the residuals, as shown in Figure 6.9.

(7) The largest residual now occurs at (3, 1), where $R_{3,1} = -265$. We therefore reduce $T_{3,1}$ by 65, with the result that

$$R_{3,1} = -265 + (4 \times 65) = -5$$
$$R_{2,1} = 10 - 65 = -55$$
$$R_{3,2} = 0 - 65 = -65$$

This change in $T_{2,3}$, and the new residuals, are entered on the diagram, as shown in Figure 6.10.

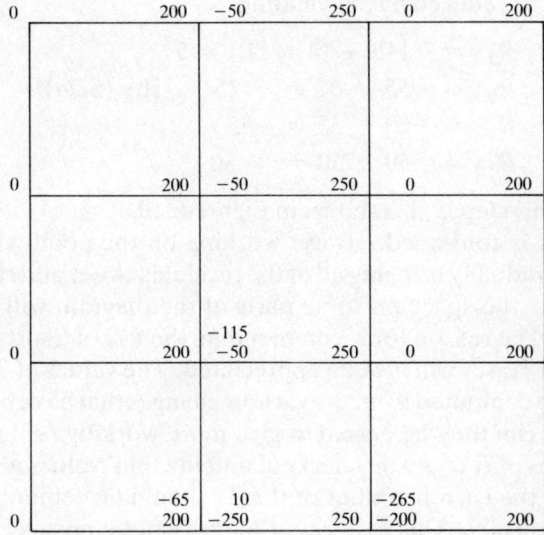

Figure 6.9 Relaxation by hand-calculation: the result of $\delta T_{2,1} = -65$.

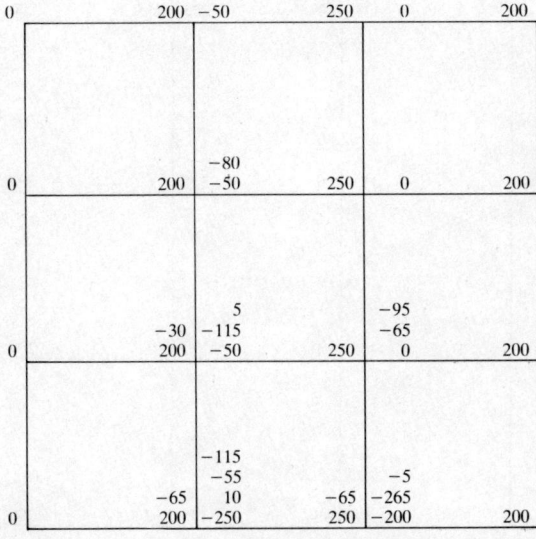

Figure 6.10 Relaxation by hand-calculation: the result of $\delta T_{3,1} = -65$ and $\delta T_{2,2} = -30$.

(8) Next, $T_{2,2}$ is reduced by 30, yielding

$$R_{2,2} = -115 + (4 \times 30) = 5$$
$$R_{2,1} = -55 - 60 = -115 \quad \text{[by (6.24)]}$$
$$R_{3,2} = -65 - 30 = -95$$
$$R_{2,3} = -50 - 30 = -80$$

The result of this step is also shown in Figure 6.10.

This process is continued, always working on the point with the largest residual, and gradually bringing all of the residuals closer and closer to zero.

After a while the space on some parts of the diagram will become filled with working. The reason for recommending the use of a soft pencil and the provision of an eraser will now be appreciated. The values of T at each mesh point should be computed from the various changes that have been recorded. These changes can then be erased to give more working room. At the same time, the values of R could be checked, and the 'old' values also erased.

In any case, the current values of the Ts should be computed and the Rs checked periodically. This is because an arithmetic mistake can easily be made. For example, suppose the stage shown in Figure 6.11 is reached. (Note that only the *current* values of T and R are shown; the previous working has been erased.) A check of the residuals shows that $R_{3,3}$ is incorrect: it should be -6 instead of $+1$. The error here is not serious – but it could be. The beauty of this method is that recovery from this error is easy.

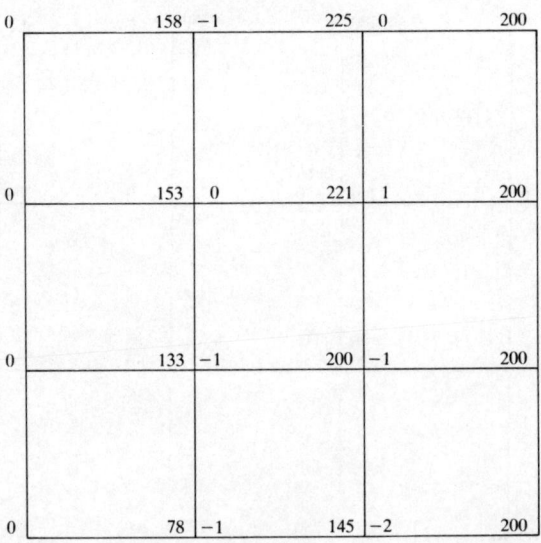

Figure 6.11 Relaxation by hand-calculation: an error in $R_{3,3}$.

NON-RECTANGULAR SOLUTION REGIONS

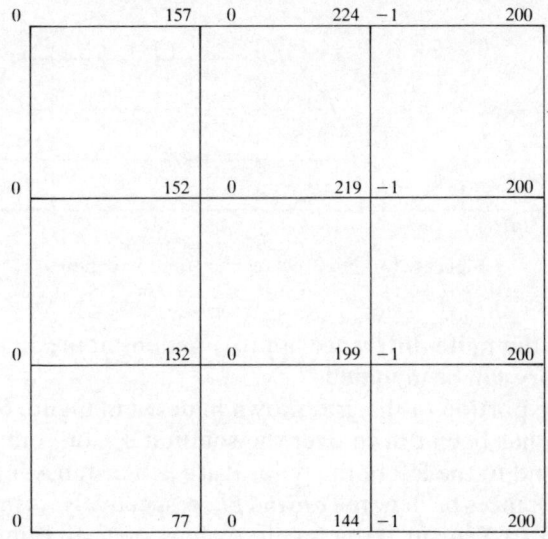

Figure 6.12 Relaxation by hand-calculation: the final answer.

It is simply necessary to enter the correct residual and continue. A few more steps leads to the situation shown in Figure 6.12.

The solution has now been carried as far as it can without going to fractional changes in T. Since a very coarse mesh has been used, the truncation errors will be large and further refinement of the values of T is not warranted.

6.10 Non-rectangular solution regions

So far, we have considered only rectangular solution regions, within which a rectangular grid such as that shown in Figure 6.2 can be readily placed. However, if a portion of the boundary is not parallel to either axis, and in particular if a portion of the boundary is not a straight line, then a rectangular grid may not fit neatly into the solution region.

For the triangular region shown in Figure 6.13a a square grid may still be used. Provided Dirichlet conditions are specified along the sloping part of the boundary, the solution can proceed as described above. If Neumann or mixed conditions apply, then the situation becomes more complex and is beyond the scope of this introductory treatment.

For the region shown in Figure 6.13b two approaches are possible. First, a *non-rectangular* grid can be constructed which is appropriate to the shape of the boundary; such an approach is again beyond the scope of this book.

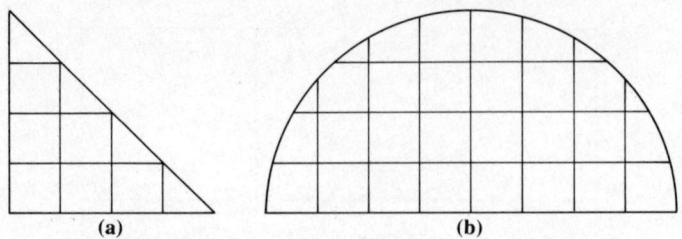

Figure 6.13 Non-rectangular solution regions.

Alternatively, the finite difference approximations at mesh points near the curved boundary can be modified.

Consider the portion of the grid shown in detail in Figure 6.14. A square mesh of size h has been drawn over the solution region, but the boundary points above and to the left of the point P are at a distance less than h from P. Let these distances be denoted αh and βh, respectively. A finite difference approximation to $\nabla^2 u$, in terms of the values of u at P and at the four surrounding mesh points, can be derived as follows.

Let a, b, c and d denote the mid-points of the line segments AP, BP, CP and DP, respectively. Then central difference approximations to the first derivatives of u at these points are

$$\left.\frac{\partial u}{\partial y}\right|_a = \frac{u_A - u_P}{\alpha h} \qquad \left.\frac{\partial u}{\partial x}\right|_b = \frac{u_P - u_B}{\beta h}$$

$$\left.\frac{\partial u}{\partial x}\right|_c = \frac{u_C - u_P}{h} \qquad \left.\frac{\partial u}{\partial y}\right|_d = \frac{u_P - u_D}{h}$$

Approximations to the second derivatives of u at P are then given by

$$\frac{\partial^2 u}{\partial x^2} = \frac{(\partial u/\partial x)_c - (\partial u/\partial x)_b}{\beta h/2 + h/2} = \frac{2}{\beta h^2(\beta + 1)}\{u_B - (\beta + 1)u_P + \beta u_C\}$$

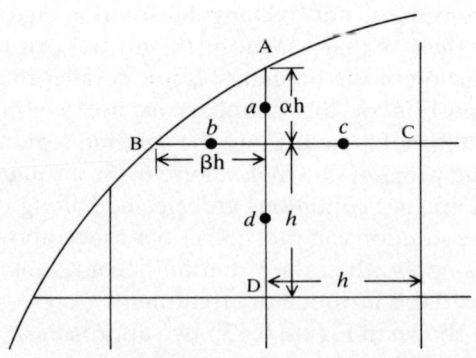

Figure 6.14 A mesh point near a curved boundary.

and

$$\frac{\partial^2 u}{\partial y^2} = \frac{(\partial u/\partial x)_a - (\partial u/\partial x)_d}{\alpha h/2 + h/2} = \frac{2}{\alpha h^2(\alpha + 1)} \{u_A - (\alpha + 1)u_P + \alpha u_D\}$$

and the Laplacian operator $\nabla^2 u$ becomes

$$\frac{\partial^2 u}{\partial x^2} + \frac{\partial^2 u}{\partial y^2} = \frac{2}{\alpha h^2(\alpha + 1)} \{u_A - (\alpha + 1)u_P + \alpha u_D\} + \frac{2}{\beta h^2(\beta + 1)} \\ \times \{u_B - (\beta + 1)u_P + \beta u_C\} \quad (6.25)$$

Note that if $\alpha = \beta = 1$, then the regular approximation in (6.9) is recovered.

Worked example

Solve Laplace's equation $\nabla^2 u = 0$ in the semicircular region of unit radius shown in Figure 6.13b, subject to the boundary conditions $u = 1$ along $y = 0$, and $u = 0$ on the curved boundary. (The conflict in boundary values at each end of the diameter does not matter since those points are not used in the calculations.)

The approximation defined in (6.25) must be extended to allow for the fact that some internal mesh points will have the curved boundary on their left (in the orientation of Fig. 6.13b) and others will have that boundary on their right. To make it completely general, it could also be made to include the possibility of a curved boundary below a mesh point, although that has not been done here. When the analysis leading to (6.25) is thus extended and the resulting equation rearranged, we obtain

$$u_{i,j} = \frac{\alpha \beta_l \beta_r}{\alpha + \beta_l \beta_r} \left(\frac{u_{i+1,j}}{\beta_l(\beta_l + \beta_r)} + \frac{u_{i-1,j}}{\beta_r(\beta_l + \beta_r)} + \frac{u_{i,j+1}}{\alpha(\alpha + 1)} + \frac{u_{i,j-1}}{\alpha + 1} \right) \quad (6.26)$$

where α is the distance (in units of the mesh size h) from any mesh point to the mesh point *above* (i.e. in the direction of increasing y) and β_l and β_r are the distances to the neighbouring points to the *left* and *right*. For most mesh points α, β_l and β_r will all equal unity, but for mesh points adjacent to the curved boundary one or two of them will be less than unity.

It is necessary to determine which mesh points are adjacent to the curved boundary, and to calculate α and the βs for those points. Figure 6.15 defines some quantities used for this purpose:

i, j mesh line numbers in the x- and y-directions, respectively;

j_{max} the value of j at the last mesh point inside the solution region for each value of i;

x_{min}, x_{max} the x co-ordinates at the boundary along the line $j = j_{max}$;

i_{min}, i_{max} the values of i at the mesh points just inside the boundary on the line $j = j_{max}$;

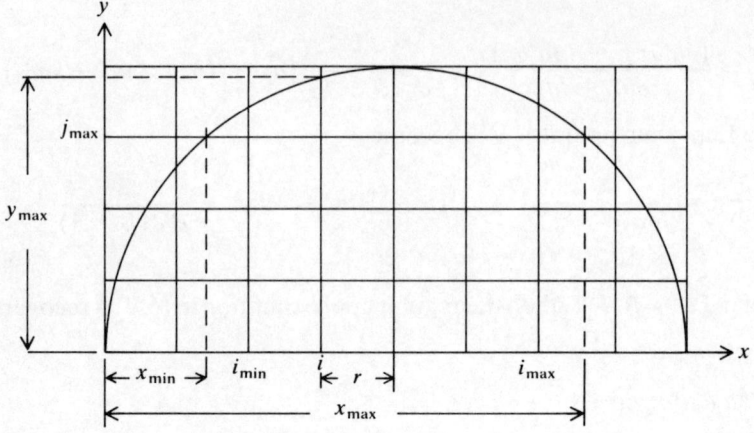

Figure 6.15 Notation for the solution of $\nabla^2 u = 0$ in a semicircle.

y_{max} the y co-ordinate at the boundary along any line i; and
r the distance from the centre of the semicircle to a mesh point on the x-axis.

From the figure it can be seen that, for each mesh point at a position x along the x-axis,

$$y_{max} = (1 - r^2)^{1/2} \quad \text{where } r = |1 - x|$$

and that for each mesh point at a position y along any mesh line $x =$ constant,

$$x_{min} = 1 - (1 - y^2)^{1/2} \quad \text{and} \quad x_{max} = 1 + (1 - y^2)^{1/2}$$

The row and column numbers defining the points just inside the curved boundary are given by

$$j_{max} = \text{integer part of } (y_{max}/h) + 1$$
$$i_{min} = \text{integer part of } (x_{min}/h) + 2$$
$$i_{max} = \text{integer part of } (x_{max}/h) + 1$$

The logic of these expressions can best be confirmed by considering particular mesh points.

Figure 6.16 shows a program to solve this problem using Gauss–Seidel iteration on the system represented by (6.26). A square mesh has been used, so there are twice as many intervals along the x-axis as along the y-axis. A rectangular array u has been dimensioned, although points outside the semicircle are neither computed nor printed.

Figure 6.16 The solution of $\nabla^2 u = 0$ in a semicircle, with $u = 0$ on the circumference and $u = 1$ on the diameter.

```
c
        implicit real*8 (a-h,o-z)
        logical conv
        dimension u(65,33)
c
   1    write(*,2)
   2    format(/' enter no of intervals on diameter, itmax and eps :  ',$)
        read(*,*) m, itmax, eps
        if (m .le. 0 .or. m .gt. 64) stop
        if (m .ne. 2*(m/2)) then
            write (*,4)
   4        format(/' please use an even number of intervals'/)
            go to 1
        endif
        n = m/2
c
c       m and n are the number of intervals along the x and y axes
c
        mp1 = m+1
        np1 = n+1
c
        mmid = np1
        nmid = 1+n/2
c
c       mmid and nmid define a point in the solution region
c       where the value of u will be monitored
c
        h = 1./n
c
c       h is the mesh size
c
c       initialise:
c
        do 5 i=1,mp1
        u(i,1) = 1.
        do 5 j=2,np1
   5    u(i,j) = 0.
        iter = 0
c
c       gauss-seidel loop starts here
c
  10    conv = .true.
        iter = iter + 1
c
c       for each interior mesh point i along the x-axis ...
c
        do 50 i = 2,m
        x = (i-1)*h
        r = dabs(1.-x)
        ymax = dsqrt(1.-r*r)
        jmax = idint(ymax/h) + 1
c
c       ... and for each interior mesh point j along the mesh
c       line through i ...
c
        do 50 j = 2,n
        y = (j-1)*h
        xmin = 1. - dsqrt(1.-y*y)
        xmax = 1. + dsqrt(1.-y*y)
        imin = idint(xmin/h) + 2
        imax = idint(xmax/h) + 1
c
c       assume alpha and both beta's are unity
c
        alpha = 1.
        betao = 1.
        betau = 1.
```

(continued)

Figure 6.16 (*continued*)

```
c
c       for points outside the semicircle, jump to the end of the loop
c
           if (i .lt. imin)    go to 50
           if (i .gt. imax)    go to 50
           if (j .gt. jmax)    go to 50
c
c       now calculate alpha and the beta's for points near the boundary
c
           if (j .eq. jmax)    alpha = (ymax - (jmax-1)*h)/h
           if (i .eq. imin)    betao = ((imin-1)*h - xmin)/h
           if (i .eq. imax)    betau = (xmax - (imax-1)*h)/h
c
           alsum = alpha + 1.
           besum = betao + betau
           beprod = betao*betau
c
           unew = (alpha*beprod/(alpha+beprod))*(
    +            u(i-1,j)/betao/besum
    +          + u(i+1,j)/betau/besum
    +          + u(i,j-1)/alsum
    +          + u(i,j+1)/alpha/alsum              )
c
           if (dabs(u(i,j)-unew) .gt. eps) conv = .false.
c
           u(i,j) = unew
c
   50   continue
c
        if (conv) then
           write (*,60) iter
   60      format(/' solution has converged in ',i4,' iterations'/)
           write (*,65) mmid, nmid, u(mmid,nmid)
   65      format(' value at point (',i2,',',i2,') = ',g11.5/)
           if (n .le. 8) then
c
c              write out the solution
c
               do 70 i=1,mp1
               x = (i-1)*h
               r = dabs(1.-x)
               ymax = dsqrt(1.-r*r)
c
c       see the text regarding the definition of jtop
c
               jtop = idint(ymax/h-0.00001) + 2
   70          write(*,80)( u(i,j), j=1,jtop)
   80          format(1p8g9.3)
           endif
           go to 1
        else if (iter .gt. itmax) then
           write (*,90) itmax, mmid, nmid, u(mmid,nmid)
   90      format(/' failed to converge in ',i4,' iterations.'/
    +              ' current value at point (',i2,',',i2,') = ',g11.5/)
           go to 1
        else
           go to 10
        endif
        end
```

Typical output from this program appears as follows:

```
enter no of intervals on diameter, itmax and eps :  8   100 0.000001
solution has converged in   27 iterations
value at point ( 5, 3) =  .40447

1.00      .000
1.00      .408     .121     .000
1.00      .591     .292     7.923E-02 .000
1.00      .663     .379     .154      .000
1.00      .683     .404     .178      .000
1.00      .663     .379     .154      .000
1.00      .591     .292     7.923E-02 .000
1.00      .408     .121     .000
1.00      .000

enter no of intervals on diameter, itmax and eps :  0   0   0.0
```

For simplicity, the results have been printed with the x-axis running down the page and the y-axis across it: in other words, Figure 6.15 has been rotated through 90° in a clockwise direction. Notice that the last values on each line of the table of output values (except for the middle line, where the last value is on the boundary) are for points *outside* the solution region; the boundary does not in general coincide with mesh points. In the section of the program where these values are printed, the variable 'jtop' has been defined in such a way that the zero value at the mesh point outside the solution region at $x = 1$ is not printed.

Solutions were computed for three values of h; the results for $u(1, \frac{1}{2})$ were:

Solution	h	u	Error
1	0.25	0.404 47	−0.005 20
2	0.125	0.408 03	−0.001 64
3	0.0625	0.409 19	−0.000 48
R_{12}		0.409 22	−0.000 45
R_{23}		0.409 58	−0.000 09

A Richardson's extrapolation on the first two values predicts 0.409 22 as the exact solution; using the second and third values, the result is 0.409 58. In fact, the exact value can be found: it is 0.409 67. It is seen that the errors reduce by a factor of a little over three for each halving of the mesh size. The factor would be four if the truncation error were exactly $O(h^2)$, and thus convergence is not quite quadratic. This deterioration of performance is not uncommon when the mesh size is not uniform throughout the whole solution region. It can also be seen that Richardson's extrapolation – especially using the two finer mesh sizes – leads to a very good answer.

6.11 Higher-order equations

So far, we have only considered second-order equations. An important fourth-order equation, which can also be said to be elliptic (in the sense that the solution region is closed and boundary conditions are specified at all points of its boundary) is the *biharmonic equation*

$$\nabla^4 u = \frac{\partial^4 u}{\partial x^4} + 2\frac{\partial^4 u}{\partial x^2 \partial y^2} + \frac{\partial^4 u}{\partial y^4} = f(x, y) \tag{6.27}$$

This equation describes, for example, the deflection of a thin plate which is fixed at its edges and subjected to a uniform pressure over its surface. It also applies to the steady, slow motion (in two dimensions) of a viscous, incompressible fluid. The boundary conditions which are most commonly encountered provide values for the function u and either its first or its second normal derivative at each boundary point.

In the notation of Chapter 4, (6.27) may be written

$$(D_x^4 + 2D_x^2 D_y^2 + D_y^4)u = f$$

From (4.35), the derivative operators (in either direction) may be written

$$h^2 D^2 \equiv \delta^2 + O(h^4) \quad \text{and} \quad h^4 D^4 \equiv \delta^4 + O(h^6)$$

and therefore (6.27) may be written

$$\frac{\delta_x^4 u + 2\delta_x^2 \delta_y^2 u + \delta_y^4 u}{h^4} = f + O(h^2) \tag{6.28}$$

where, for simplicity, the same mesh size h has been assumed in the x- and y-directions.

The fourth central difference (in the x-direction, for example) is

$$\delta_x^4 u_{i,j} = u_{i+2,j} - 4u_{i+1,j} + 6u_{i,j} - 4u_{i-1,j} + u_{i-2,j}$$

The *mixed* derivative is obtained by applying the second central difference operators in succession:

$$\begin{aligned}\delta_x^2 \delta_y^2 u_{i,j} &= \delta_x^2 (u_{i,j+1} - 2u_{i,j} + u_{i,j-1}) \\ &= u_{i+1,j+1} - 2u_{i,j+1} + u_{i-1,j+1} \\ &\quad -2(u_{i+1,j} - 2u_{i,j} + u_{i-1,j}) \\ &\quad + u_{i+1,j-1} - 2u_{i,j-1} + u_{i-1,j-1}\end{aligned}$$

The biharmonic operator ∇^4 therefore yields an expression involving values of u at 13 mesh points:

$$\begin{aligned}\nabla^4 u = \quad & u_{i-2,j} \\ & + 2u_{i-1,j-1} - 8u_{i-1,j} + 2u_{i-1,j+1} \\ + u_{i,j-2} & - 8u_{i,j-1} + 20u_{i,j} - 8u_{i,j+1} + u_{i,j+2} \\ & + 2u_{i+1,j-1} - 8u_{i+1,j} + 2u_{i+1,j+1} \\ & + u_{i+2,j}\end{aligned} \tag{6.29}$$

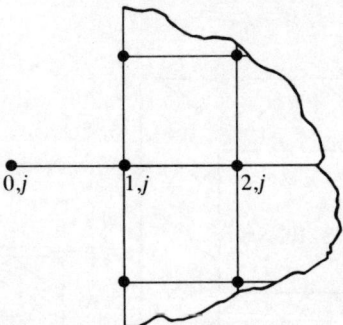

Figure 6.17 A fictitious mesh point outside a boundary.

Equation (6.29) requires modification when it is to be applied at mesh points which are adjacent to a boundary, since one (or, near a corner, two) of the values needed is at points outside the solution region. This modification is made using the boundary conditions.

Consider the portion of a solution region shown in Figure 6.17, and suppose that u and $\partial u/\partial x$ are known along the boundary. Introducing a fictitious mesh point at $(0, j)$ outside the solution region, as shown, the value of u there is given by the derivative boundary condition:

$$u_{0,j} = u_{2,j} - 2h \left. \frac{\partial u}{\partial x} \right|_{1,j}$$

and this expression can be inserted into (6.29) as required. A similar strategy can be employed when the second normal derivative is given as a boundary condition.

Problems

1. Figure 6.18 shows two views of a hydrostatic bearing pad. Pressurized oil is introduced into the central region at a pressure $p_s = 50 \times 10^4$ N m^{-2} gauge, and exhausts to the atmosphere at all edges. At any point in the hatched region the pressure distribution is given by the solution of

$$\frac{\partial^2 p}{\partial x^2} + \frac{\partial^2 p}{\partial y^2} = 0$$

(a) Using a square mesh with sides 100 mm, find the pressure at all points using Gauss–Seidel iteration. Take $p = 30 \times 10^4$ N m^{-2} as starting values.

(b) Suggest how to increase the rate of convergence.

(c) Write a computer program to solve the same problem. Use a range of values of h, together with Richardson's extrapolation, until your values for p are accurate to three significant figures.

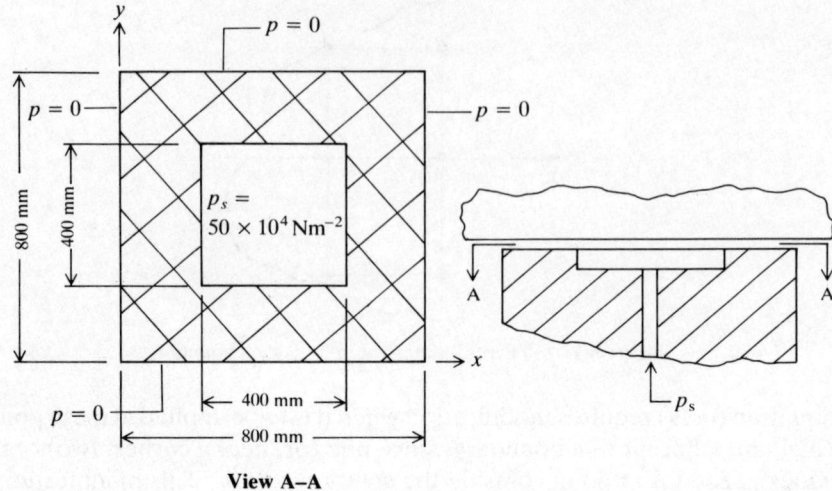

Figure 6.18 Problem 1.

2. The stress due to torsion in a solid elastic cylinder of square cross section defined by the lines $x = \pm 1$, $y = \pm 1$ is given by the solution of

$$\frac{\partial^2 u}{\partial x^2} + \frac{\partial^2 u}{\partial y^2} + 2 = 0$$

subject to the boundary condition $u = 0$ on all surfaces.

(a) Using a square mesh with $h = \frac{1}{2}$, find the stress at all internal mesh points.

(b) Write a computer program to solve the same problem. Use a range of values of h, together with Richardson's extrapolation, until your values for u are accurate to three significant figures.

(c) Compare your solutions at the origin with the analytical solution, which is 0.589.

3. Repeat parts (a) and (b) of Problem 2 for the hollow cylinder of which a portion is shown in Figure 6.19. The cross section is symmetrical about vertical and horizontal lines.

4. Repeat Problem 3 for the triangular cylinder shown in Figure 6.20.

5. Modify the program of Figure 6.16 to solve the torsion equation of Problem 2 in the semicircular region shown in Figure 6.15.

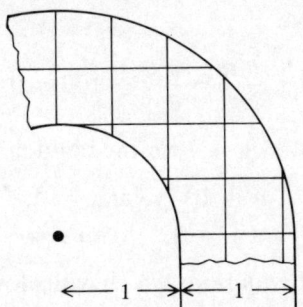

Figure 6.19 Problem 3.

6. The (non-dimensional) temperature θ in a thin metal plate subject to specified boundary conditions at its edges and also exchanging heat between its surface and the surroundings can be given by the solution of

$$\frac{\partial^2 \theta}{\partial x^2} + \frac{\partial^2 \theta}{\partial y^2} + k\theta = 0$$

where k is a parameter depending on the surface heat transfer coefficient and on the method of making the equation non-dimensional.

(a) How does the choice of k affect the numerical solution? Consider in particular (i) $k \leq 0$ and (ii) $k = 4/h^2$, where h is the mesh size. What is the range of values of k for which iterative methods of solution will fail?

(b) Solve the problem (on a computer) for $k = -10$ and the boundary conditions

$$\theta = y \quad \text{at } x = 0 \qquad \theta = y^2/2 \quad \text{at } x = 1$$
$$\theta = 0 \quad \text{at } y = 0 \qquad \partial\theta/\partial y = 1 \quad \text{at } y = 1$$

Use a range of values of h, together with Richardson's extrapolation, until your values for θ are accurate to three significant figures.

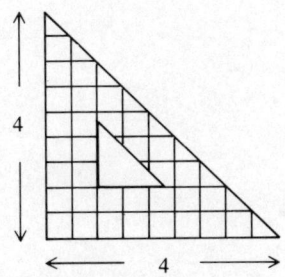

Figure 6.20 Problem 4.

7. It is desired to solve

$$\frac{\partial^2 u}{\partial x^2} + \frac{\partial^2 u}{\partial y^2} + 160 = 0$$

in the square region $0 \leq x, y \leq 1$ for the boundary conditions

$$u = 10 \quad \text{at } x = 0, 1$$
$$\partial u/\partial y = 100 \quad \text{at } y = 0 \qquad \partial u/\partial y = -100 \quad \text{at } y = 1$$

(a) Set up a system of equations which will allow the solution to be found for $h = 0.25$, and perform two or three iterations of the solution of the system.

(b) Write a computer program to solve the same problem. Use a range of values of h, together with Richardson's extrapolation, until your values for u are accurate to three significant figures.

7
Partial differential equations II – parabolic equations

7.1 Introduction

The simplest parabolic differential equation is (6.3) or, as we normally meet it, (6.3a):

$$\frac{\partial u}{\partial t} = a \frac{\partial^2 u}{\partial x^2} \qquad (7.1)$$

in which u is given as a function of a space variable x and of time t, and in which the coefficient a is a constant. Strictly, this is a two-dimensional equation: u is a function of two co-ordinates. However, in the form, for example, of (7.2) below, this is usually called the 'one-dimensional heat conduction equation' because it describes the unsteady conduction of heat in one *space* dimension.

We shall consider several simple finite difference approximations to equations like (7.1), and study methods for their solution. We shall also study the convergence and stability of these approximations – in effect, how accurate they are, and whether in fact they can be used.

7.2 The conduction equation

To fix our ideas, let us consider the one-dimensional unsteady flow of heat through a slab of material of constant and uniform thermal diffusivity* α and thickness L. The equation describing the distribution of temperature $T(X, \tau)$ in space and time is

$$\frac{\partial T}{\partial \tau} = \alpha \frac{\partial^2 T}{\partial X^2} \qquad (7.2)$$

---

* The thermal diffusivity α of a substance is given by $\alpha = K/\rho C$, where K, ρ and C are the thermal conductivity, density and specific heat, respectively, of the substance. Its units are $m^2 s^{-1}$.

where T, τ and X denote, respectively, the dimensional temperature, time and position. From a mathematical viewpoint three auxiliary conditions are required for the full solution of this equation, giving T or its first (spatial) derivative at one value of τ (for all values of X) and at one or two values of X (for all τ greater than the initial value). From a physical viewpoint the commonest conditions are probably a specified temperature distribution at a time which can be designated $\tau = 0$, and specified values of temperature or heat flux at the two faces of the material $X = 0$ and $X = L$. We will therefore consider the conditions

$$T(X, 0) = T_i(X) \quad \text{for } 0 \leq X \leq L, \tau = 0$$
$$T(0, \tau) = T_a(\tau) \quad \text{for } \tau > 0, X = 0 \quad (7.3)$$
$$T(L, \tau) = T_b(\tau) \quad \text{for } \tau > 0, X = L$$

The first of these is known as the *initial* condition, the others are *boundary* conditions. For generality, T_i is permitted to be a function of position, and T_a and T_b to be functions of time.

7.3 Non-dimensional equations yet again

The first task is to generalize the results which we will, in due course, obtain by making the variables *non-dimensional*. The method of doing this will be suggested by the problem itself, and it is usually easy to find reference quantities for temperature and length. In the present example we can make the substitutions

$$\theta = T/T^* \quad x = X/L$$

where T^* is a known reference temperature. For example, we could choose $T(0, 0)$ for T^*, but $T(0, L)$ would be just as good a choice. The choice of L as the reference length is obvious: indeed, there is only one length available to use, and it is suitable because it leads to a new, dimensionless variable x which lies in value between 0 and 1. However, there is no obvious 'characteristic time scale' to use as a reference time, so tentatively we define a dimensionless time by

$$t = \tau/\tau^*$$

where τ^* is not yet specified.

To transform (7.2) into the new, non-dimensional variables, we note that

$$\frac{\partial T}{\partial \tau} = \frac{\partial (T^*\theta)}{\partial t} \frac{dt}{d\tau} = \frac{T^*}{\tau^*} \frac{\partial \theta}{\partial t}$$

Similarly,

$$\frac{\partial T}{\partial X} = \frac{\partial (T^*\theta)}{\partial x} \frac{dx}{dX} = \frac{T^*}{L} \frac{\partial \theta}{\partial x} \quad \text{and} \quad \frac{\partial^2 T}{\partial X^2} = \frac{T^*}{L^2} \frac{\partial^2 \theta}{\partial x^2}$$

Then (7.2) becomes

$$\frac{T^*}{\tau^*}\frac{\partial \theta}{\partial t} = \frac{\alpha T^*}{L^2}\frac{\partial^2 \theta}{\partial x^2}$$

The choice of

$$\tau^* = L^2/\alpha$$

now suggests itself, because it will result in all the parameters of the problem cancelling, yielding

$$\frac{\partial \theta}{\partial t} = \frac{\partial^2 \theta}{\partial x^2} \tag{7.4}$$

The initial and boundary conditions become

$$\theta(x, 0) = \theta_i(x) \quad \text{for } 0 \leq x \leq 1, t = 0 \tag{7.5a}$$

$$\theta(0, t) = \theta_a(t) \quad \text{for } t > 0, x = 0 \tag{7.5b}$$

$$\theta(1, t) = \theta_b(t) \quad \text{for } t > 0, x = 1 \tag{7.5c}$$

Equations (7.4) and (7.5) describe, in dimensionless variables, the problem we wish to solve.

There is another way of defining a dimensionless temperature which can be used if there are *two* specified temperatures, T_1 and T_2, connected with the problem being solved. For example, $T_a(0)$ and $T_b(0)$ may be used. It then proves convenient to define θ as

$$\theta = \frac{T - T_2}{T_1 - T_2} \tag{7.6}$$

The advantage of this definition is that if T_1 and T_2 are, respectively, the maximum and minimum temperatures experienced by the material, then $0 \leq \theta \leq 1$. Computationally, it is a good idea to keep the values of the unknown quantity bounded, and this range for θ means that erroneous values, outside the range, are readily detected. Equation (7.4) is unaffected by the use of this alternative definition of θ.

7.4 Notation

Equations (7.4) and (7.5) can be solved analytically, but they are nevertheless convenient to illustrate the numerical techniques to be used on more difficult problems. Instead of seeking an analytical expression for $\theta(x, t)$ which would allow us to evaluate the dimensionless temperature at *any* position or time, we shall replace the differential equation by a finite difference approximation. This will comprise a system of algebraic equations, the solutions of which will (or at least should) be a good approximation

246   PARTIAL DIFFERENTIAL EQUATIONS II

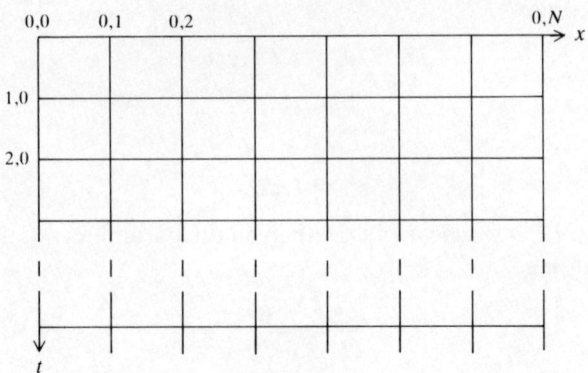

Figure 7.1 Mesh notation for the solution of a parabolic equation.

to the solution of the original differential equation. In a manner similar to that used for elliptic equations in Chapter 6, we shall look for numerical values of θ at the nodes or mesh points shown in Figure 7.1.

We adopt the notation

$$\theta(x, t) \equiv \theta(j\Delta x, m\Delta t) \equiv \theta(j, m) \equiv \theta_{j,m}$$

where Δx and Δt are the mesh sizes in space and time, respectively and j and m are integers, j taking values between 0 and $n = 1/\Delta x$, and m taking values increasing from zero. The time axis has been drawn downwards in Figure 7.1 to match the tables (such as Table 7.1, below), which contain example solutions.

7.5 An explicit method

In seeking a finite difference approximation to (7.4) we recall two expressions for numerical differentiation derived previously, namely

$$D\theta = \frac{1}{h}(\Delta - \Delta^2/2 + \Delta^3/3 - \Delta^4/4 + \ldots)\theta \qquad (4.27)$$

and

$$D^2\theta = \frac{1}{h^2}\{\delta^2 - (1/12)\delta^4 + (1/90)\delta^6 - \ldots\}\theta \qquad (4.41)$$

where h is the step size in the variable with respect to which θ is being differentiated. Equation (4.27) will be used to approximate $\partial\theta/\partial t$, and for brevity Δt, the step size in time, will be denoted by k. Retaining only the first term on the right of (4.27), we obtain

$$\frac{\partial \theta}{\partial t} = \frac{\Delta \theta}{k} + O(k) = \frac{\theta_{j,m+1} - \theta_{j,m}}{k} + O(k) \qquad (7.7)$$

AN EXPLICIT METHOD

Similarly, the use of one term from (4.41) yields

$$\frac{\partial^2 \theta}{\partial x^2} = \frac{\delta^2 \theta}{h^2} + O(h^2) = \frac{\theta_{j+1,m} - 2\theta_{j,m} + \theta_{j,m-1}}{h^2} + O(h^2) \qquad (7.8)$$

where h is the spatial step size Δx.

We substitute (7.7) and (7.8) into (7.4), neglecting the truncation errors, to obtain

$$\frac{\theta_{j,m+1} - \theta_{j,m}}{k} = \frac{\theta_{j+1,m} - 2\theta_{j,m} + \theta_{j-1,m}}{h^2} \qquad (7.9)$$

which may be written

$$\theta_{j,m+1} = \theta_{j,m} + r(\theta_{j+1,m} - 2\theta_{j,m} + \theta_{j-1,m}) \qquad (7.10)$$

where $r = k/h^2 = \Delta t/\Delta x^2$.

This equation defines θ at the point j and at the time $(m + 1)$ in terms of three values of θ at the previous time m. Since the initial condition defines θ for all values of x at $t = m = 0$, θ may be obtained from (7.10) at time $m = 1$ for each value of j between (and including) 1 and $(n - 1)$. Equation (7.10) *cannot* be used at $j = 0$ or $j = n$, because values of θ at points outside the solution region, at $j = -1$ and $j = n + 1$, would be invoked. However, because of the boundary conditions (7.5b and c), values of θ at these boundary points are known at all times, and therefore there is no need for them to be calculated.

The values of θ at $m = 1$ may be calculated in any order, although for simplicity in programming they would normally be calculated in the order $j = 1, j = 2, \ldots, j = n - 1$.

Once the interior values at time $m = 1$ have been computed, and since the values of $\theta(1, 0)$ and $\theta(1, n)$ are given by the boundary conditions, all values of θ at $m = 1$ are known. Values of $\theta(2, j)$ may now be obtained at all of the internal points $j = 1, j = 2$, etc. When the calculations at $m = 2$ are complete we can move on to $m = 3$. We can thus work our way steadily along the time axis, from $m = 1$ to $m = 2$ to $m = 3$, and so on. This process is sometimes called a *marching* solution.

The unknown $\theta_{j,m+1}$ is given *explicitly* by (7.10); the right-hand side contains no unknown quantities. For this reason, the process is also called an *explicit* method.

Finally, because it uses a forward difference for the time derivative and a central difference for the space derivative, it is also known as the forward time, central space (FTCS) scheme.

It should be remembered that when (7.10) was constructed the truncation error terms were neglected. The solution of (7.10) must therefore be different from the solution of (7.4). We hope that the difference would be reduced if the mesh sizes were to be reduced, and we shall discuss this question of *convergence* in a later section. To avoid complicating the notation, we will generally use the same symbol (here we are using θ) for the solutions of both the differential equation and the difference equation. It will be clear from the context which is intended – and it should always be remembered that the two are not quite the same.

Worked example 1

Suppose we have a concrete wall 0.5 m thick, of which the thermal diffusivity is 1.25×10^{-5} m^2 s^{-1}. Its temperature at a given time is uniform at 20°C. Let us imagine that the temperature of one surface of the wall – the surface at $x = 0$ – is suddenly raised to 80°C and both surface temperatures remain fixed thereafter. What is the temperature distribution throughout the slab after 1 h?

If we use (7.6) to define θ, then $T_1 = 80$, $T_2 = 20$ and

$$\theta = \frac{T - 20}{60}$$

The initial and boundary conditions then become

$$\begin{aligned}
\theta(x, 0) &= 0 \quad \text{for } 0 \leq x \leq 1, t = 0 \\
\theta(0, t) &= 1 \quad \text{for } t > 0, x = 0 \\
\theta(1, t) &= 0 \quad \text{for } t > 0, x = 1
\end{aligned}$$

We have an immediate difficulty. There is a discontinuity in temperature at $x = 0$ and $t = 0$, i.e. at the origin in Figure 7.1. Should the temperature there be taken as 0 or 1? In a real situation there cannot be an instantaneous change in temperature – no matter how rapidly the heating occurs, it will always require a finite time for the change from 20°C to 80°C to occur. There is no 'best' way out of this difficulty, in the sense of making a best match to the physical situation, since that itself is vague. Perhaps the most appropriate step – but one which we can only take intuitively – is to set $\theta(0, 0) = 0.5$ and $\theta(0, t) = 1$ for all $t > 0$.

To reduce the amount of calculation, we will choose $h = 0.2$, and we will set $r = 0.5$, in which case (7.10) takes the particularly simple form

$$\theta_{j,m+1} = \tfrac{1}{2}(\theta_{j+1,m} + \theta_{j-1,m}) \tag{7.11}$$

The temperature at each internal mesh point and at any time is thus, for $r = 0.5$, the average of the temperatures at the two neighbouring mesh points at the previous time.

With $h = 0.2$ and $r = 0.5$, we have that $k = rh^2 = 0.02$. In terms of the dimensional variables the space mesh size is 0.1 m and the time step is of duration

$$\Delta\tau = \tau^*\Delta t = (L^2/\alpha)\Delta t = (0.5^2/1.25 \times 10^{-5})(0.02) = 400 \text{ s}.$$

Table 7.1 The explicit solution with $r = 0.5$.

x		0	0.2	0.4	0.6	0.8	1.0
X (m)		0	0.1	0.2	0.3	0.4	0.5
τ (s)	t						
0	0.00	0.5	0.0	0.0	0.0	0.0	0.0
400	0.02	1.0	**0.25**	**0.0**	**0.0**	**0.0**	0.0
800	0.04	1.0	**0.5**	**0.125**	**0.0**	**0.0**	0.0
1200	0.06	1.0	**0.5625**	**0.25**	**0.0625**	**0.0**	0.0
1600	0.08	1.0	**0.625**	**0.3125**	**0.125**	**0.0312**	0.0
2000	0.10	1.0	**0.6563**	**0.375**	**0.1719**	**0.0625**	0.0
2400	0.12	1.0	**0.6875**	**0.4141**	**0.2188**	**0.0865**	0.0
2800	0.14	1.0	**0.7071**	**0.4532**	**0.2503**	**0.4094**	0.0
3200	0.16	1.0	**0.7266**	**0.4787**	**0.4313**	**0.1252**	0.0
3600	0.18	1.0	**0.7394**	**0.5790**	**0.3020**	**0.2157**	0.0

It is therefore necessary to perform the calculations for nine time steps to reach $\tau = 1$ h.

The values in Table 7.1 may now be calculated using (7.11). Only the interior values (those shown in **bold type**) are computed. The row immediately above, and the columns immediately to the left and right, are the given initial and boundary values. The student should verify the calculations. For example, at $x = 0.2$ and $\tau = 0.02$ (the first point at which a calculation is to be made), $\theta = \frac{1}{2}(0.5 + 0.0) = 0.25$.

The temperature distributions, at several values of τ, are shown in Figure 7.2. It is apparent that these distributions, while only approximations, have the form which would intuitively be expected.

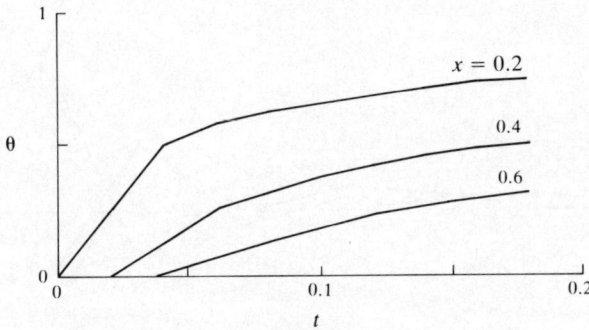

Figure 7.2 The growth of temperature with time at different positions through the wall, using $r = 0.5$.

250 PARTIAL DIFFERENTIAL EQUATIONS II

Worked example 2

The values of θ are only approximations, since the finite difference approximations (7.7) and (7.8) involve truncation errors which depend on the step sizes h and k. We might therefore expect that improved accuracy would be obtained by reducing the step sizes.

Table 7.2 The explicit solution with the mesh sizes halved.

x		0	0.1	0.2	0.3	0.4	---	0.9	1.0
X (m)		0	0.05	0.1	0.15	0.2	---	0.45	0.5
τ (s)	t								
0	0.00	0.5	0.0	0.0	0.0	0.0	---	0.0	0.0
200	0.01	1.0	0.5	0.0	0.0	0.0		0.0	0.0
400	0.02	1.0	0.5	0.5	0.0	0.0		0.0	0.0
600	0.03	1.0	1.0	0.0	0.5	0.0		0.0	0.0
800	0.04	1.0	0.0	1.5	−0.5	0.5		0.0	0.0
1000	0.05	1.0	2.5	−2.0	2.5	−1.0		0.0	0.0
1200	0.06	1.0	−3.5	7.0	−5.5	3.5	---	0.0	0.0

To test this supposition, we shall repeat the calculation with both h and k reduced to half their previous values, i.e. $h = 0.1$ and $k = 0.01$, which leads to $r = k/h^2 = 1$. Equation (7.10) then becomes

$$\theta_{j,m+1} = (\theta_{j+1,m} - \theta_{j,m} + \theta_{j-1,m}) \tag{7.12}$$

while $\Delta X = 0.05$ m and $\Delta \tau = 200$ s. The solution now starts as shown in Table 7.2 (some of the calculations have been omitted to save space) and Figure 7.3.

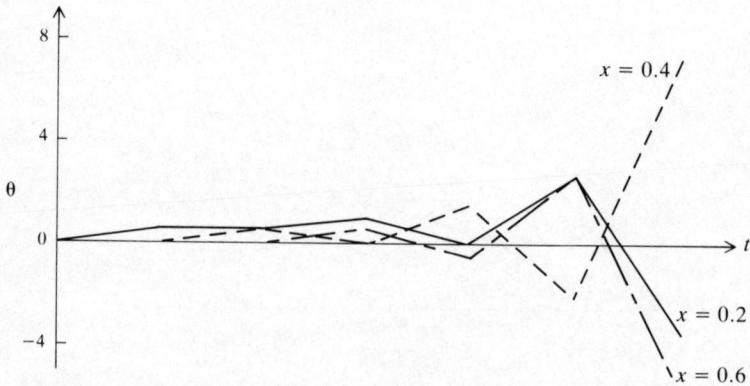

Figure 7.3 The growth of temperature with time at different positions through the wall, using $r = 1$.

Disaster has struck! Not only does Figure 7.3 look worse than Figure 7.2, rather than better, but the calculated temperatures are physically impossible, being sometimes greater than the maximum, and sometimes less than the minimum which can occur.

We have a problem. On the one hand, the truncation error terms in (7.7) and (7.8) show that the errors should diminish as h and k are reduced. On the other hand, we have the empirical fact that halving the step sizes in this case led to a catastrophic increase in the error of the solution.

This classic illustration leads us to a consideration of the *consistency*, *convergence* and *stability* of numerical solution methods.

7.6 Consistency

The consistency of a method is concerned with the accuracy with which the finite difference approximation represents the original differential equation. Truncation errors have been made. We retained only one term on the right-hand sides of (4.27) and (4.41) in developing (7.10). The extent to which these errors are significant can be found by inverting the truncated forms of (4.27) and (4.41) to find expressions, in terms of derivatives, for the finite difference expressions appearing in (7.10). We must express Δ_t and δ_x^2 in terms of D_t and D_x, respectively*.

Now,
$$\Delta_t \equiv e^{kD_t} - 1 \equiv kD_t + \frac{k^2 D_t^2}{2} + O(k^3)$$

Therefore
$$\Delta_t \theta_{j,m} = \theta_{j,m+1} - \theta_{j,m} = k\frac{\partial \theta}{\partial t} + \frac{k^2}{2}\frac{\partial^2 \theta}{\partial t^2} + O(k^3) \qquad (7.13)$$

in which the derivatives are to be evaluated at the point (j, m).

$$\delta_x^2 \equiv (2 \sinh \tfrac{1}{2} h D_x)^2 \equiv h^2 D_x^2 + \frac{h^4 D_x^4}{12} + O(h^6)$$

Therefore,
$$\delta_x^2 \theta_{j,m} = \theta_{j+1,m} - 2\theta_{j,m} + \theta_{j-1,m} = h^2 \frac{\partial^2 \theta}{\partial x^2} + \frac{h^4}{12}\frac{\partial^4 \theta}{\partial x^4} + O(h^6) \qquad (7.14)$$

Equations (7.9), (7.13) and (7.14) are now combined to obtain
$$\frac{\partial \theta}{\partial t} = \frac{\partial^2 \theta}{\partial x^2} - \frac{k}{2}\frac{\partial^2 \theta}{\partial t^2} + \frac{h^2}{12}\frac{\partial^4 \theta}{\partial x^4} + O(k^2) + O(h^4) \qquad (7.15)$$

---

* Subscripts are used here to indicate the variable with respect to which the differences or derivatives are to be calculated.

Equation (7.15) is the equivalent, in terms of derivatives, of the finite difference approximation (7.9). It enables us to see what happens to the FDA as the step sizes are reduced.

As h and k go to zero, all the terms in (7.15) after the first term on the right-hand side also go to zero, and the original differential equation (7.4) is recovered. Therefore, as h and k approach zero, the finite difference approximation (7.10) approaches the differential equation (7.4) from which it was derived. We say that (7.10) is *unconditionally consistent* with (7.4).

7.7 The Dufort–Frankel method

The unconditional consistency of (7.10) is not always achieved. This can be shown by considering another differencing scheme, known as the Dufort–Frankel method. In this method, (7.10) is replaced by

$$\theta_{j,m+1} - \theta_{j,m-1} = 2r(\theta_{j+1,m} - \theta_{j,m+1} - \theta_{j,m-1} + \theta_{j-1,m}) \quad (7.16)$$

in which the central difference approximation

$$\frac{\theta_{j,m+1} - \theta_{j,m-1}}{2k}$$

has been made for $\partial\theta/\partial t$, and in which the term $2\theta_{j,m}$ in (7.10) has been replaced by

$$\theta_{j,m+1} + \theta_{j,m-1}$$

Equation (7.16) is not self-starting; information (i.e. values of θ) at both $m = 0$ and $m = 1$ are needed before θ at $m = 2$ can be found. The initial condition will provide values at $m = 0$; those at $m = 1$ must be found by a single-step method such as FTCS.

To examine the consistency of (7.16) we again replace the differences by expressions involving derivatives. For the left-hand side we have

$$\begin{aligned}
\theta_{j,m+1} - \theta_{j,m-1} &= (E_t - E_t^{-1})\theta_{j,m} \\
&= (e^{kD_t} - e^{-kD_t})\theta_{j,m} \\
&= \left\{2kD_t + \frac{k^3}{3}D_t^3 + O(k^5)\right\}\theta_{j,m} \\
&= 2k\frac{\partial\theta}{\partial t} + \frac{k^3}{3}\frac{\partial^3\theta}{\partial t^3} + O(k^5) \quad (7.17)
\end{aligned}$$

On the right-hand side of (7.16) we first note that

$$\begin{aligned}
\theta_{j+1,m} + \theta_{j-1,m} &= (E_x + E_x^{-1})\theta_{j,m} \\
&= (e^{hD_x} + e^{-hD_x})\theta_{j,m} \\
&= 2 + h^2\frac{\partial^2\theta}{\partial x^2} + \frac{h^4}{12}\frac{\partial^4\theta}{\partial x^4} + O(h^6)
\end{aligned}$$

and, similarly, that

$$\theta_{j,m+1} + \theta_{j,m-1} = 2 + k^2 \frac{\partial^2 \theta}{\partial t^2} + \frac{k^4}{12} \frac{\partial^4 \theta}{\partial t^4} + O(k^6)$$

Therefore,

$$\theta_{j+1,m} - \theta_{j,m+1} - \theta_{j,m-1} + \theta_{j-1,m}$$
$$= h^2 \frac{\partial^2 \theta}{\partial x^2} + \frac{h^4}{12} \frac{\partial^4 \theta}{\partial x^4} - k^2 \frac{\partial^2 \theta}{\partial t^2} - \frac{k^4}{12} \frac{\partial^4 \theta}{\partial t^4} + O(h^6) + O(k^6)$$
$$= \frac{k}{r} \frac{\partial^2 \theta}{\partial x^2} + \frac{k}{r} \frac{h^2}{12} \frac{\partial^4 \theta}{\partial x^4} - k^2 \frac{\partial^2 \theta}{\partial t^2} - \frac{k^4}{12} \frac{\partial^4 \theta}{\partial t^4} + O(h^6) + O(k^6)$$

since $r = k/h^2$.

Equation (7.16) therefore becomes

$$\frac{\partial \theta}{\partial t} + \frac{k^2}{6} \frac{\partial^3 \theta}{\partial t^3} + \cdots = \frac{\partial^2 \theta}{\partial x^2} - rk \frac{\partial^2 \theta}{\partial t^2} + \frac{h^2}{12} \frac{\partial^4 \theta}{\partial x^4} - \frac{rk^3}{12} \frac{\partial^4 \theta}{\partial t^4} + \cdots$$

or, replacing rk by $(k/h)^2$,

$$\frac{\partial \theta}{\partial t} = \frac{\partial^2 \theta}{\partial x^2} - \left(\frac{k}{h}\right)^2 \frac{\partial^2 \theta}{\partial t^2} + O(h^2) + O(k^2)$$

It is therefore clear that (7.16) will approach (7.4) as h and k approach zero *only if* the ratio k/h also tends to zero. Thus, (7.16) is said to be *conditionally consistent* with (7.4).

Notice that if h and k approach zero in such a manner that the ratio k/h is a constant, Γ say, then (7.16) is consistent with the hyperbolic* equation

$$\frac{\partial \theta}{\partial t} + \Gamma^2 \frac{\partial^2 \theta}{\partial t^2} = \frac{\partial^2 \theta}{\partial x^2}$$

If h and k approach zero in this manner, then (7.16) will generate the solution of a completely different problem!

The advantage of the Dufort–Frankel method is that the transient solution is obtained more accurately; the truncation error is $O(k^2)$ compared with $O(k)$ for the FTCS method.

7.8 Convergence

We have seen that care must be taken to ensure that the difference approximation accurately represents the differential equation as the mesh sizes are reduced. However, this is not the same as ensuring that the *solution* of the difference approximation approaches the *solution* of the differential

---
* In the notation of Section 6.1, $a = 1$, $b = 0$ and $c = -\Gamma^2$. Therefore $b^2 - 4ac > 0$ and the equation is hyperbolic.

equation. We have already seen, in Table 7.2, that one does not necessarily imply the other. Although the use of $h = 0.1$ and $k = 0.01$ reduces the truncation errors of (7.10) below those associated with the use of $h = 0.2$ and $k = 0.02$, the solution of (7.4) is much less accurately represented by the solution of the finite difference approximation with the smaller mesh sizes.

A difference scheme is said to be *convergent* if its solution approaches that of the corresponding differential equation as the mesh sizes tend to zero. As with consistency, we may have *conditional* or *unconditional* convergence.

The distinction between consistency and convergence should be clearly understood. Consistency ensures, in general terms, that the *differential equation* is being properly approximated. Convergence ensures that the *solution* is being properly approximated. As (7.15) shows, (7.10) is unconditionally consistent with (7.4). However, Tables 7.1 and 7.2 show that (7.10) is, at best, only conditionally convergent.

The study of the convergence of a difference scheme, and the determination of the conditions, if any, which are necessary and sufficient to achieve convergence, are in general not easy. Fortunately, this is not serious for at least some classes of parabolic equations, as we shall shortly see. Nevertheless, the explicit difference scheme (7.10) can be analysed fairly easily.

We now need to distinguish between the exact solution of the differential equation, which we will denote by $\Theta(x, t)$, and the exact solution of the difference equation (7.10) which we will continue to call $\theta(x, t)$. The error

$$e(x, t) = \Theta(x, t) - \theta(x, t)$$

is required to vanish everywhere as $h, k \to 0$ if convergence is to be achieved. In subscript notation, we may write the solution of the difference equation as

$$\theta_{j,m} = \Theta_{j,m} - e_{j,m}$$

Substitution into (7.10) leads to

$$e_{j,m+1} = e_{j,m} + r(e_{j+1,m} - 2e_{j,m} + e_{j-1,m}) + \Theta_{j,m+1} - \Theta_{j,m} \\ - r(\Theta_{j+1,m} - 2\Theta_{j,m} + \Theta_{j-1,m}) \qquad (7.18)$$

Now, by Taylor's theorem

$$\Theta_{j,m+1} = \Theta_{j,m} + k \left(\frac{\partial \Theta}{\partial t}\right)_{x,t_1}$$

where the derivative is to be evaluated at some time t_1 such that $t \leq t_1 \leq t + k$. Similarly,

$$\Theta_{j+1,m} = \Theta_{j,m} + h \frac{\partial \Theta}{\partial x} + \frac{h^2}{2}\left(\frac{\partial^2 \Theta}{\partial x^2}\right)_{x_1,t}$$

and

$$\Theta_{j-1,m} = \Theta_{j,m} - h \frac{\partial \Theta}{\partial x} + \frac{h^2}{2}\left(\frac{\partial^2 \Theta}{\partial x^2}\right)_{x_2,t}$$

where the unsubscripted derivatives are evaluated at (x, t) and the subscripted derivatives are evaluated at (x_1, t) and (x_2, t), where $x \leq x_1 \leq x + h$ and $x - h \leq x_2 \leq x$. Substitution into (7.18) yields

$$e_{j,m+1} = e_{j,m} + r(e_{j+1,m} - 2e_{j,m} + e_{j-1,m}) + k\left(\frac{\partial \Theta}{\partial t}\right)_{x,t_1}$$
$$- \frac{rh^2}{2}\left\{\left(\frac{\partial^2 \Theta}{\partial x^2}\right) + \left(\frac{\partial^2 \Theta}{\partial x^2}\right)_{x_2,t}\right\}$$

which can be written, using $r = k/h^2$, as

$$e_{j,m+1} = (1 - 2r)e_{j,m} + r(e_{j+1,m} + e_{j-1,m}) + k\left\{\left(\frac{\partial \Theta}{\partial t}\right)_{x,t_1} - \left(\frac{\partial^2 \Theta}{\partial x^2}\right)_{x_3,t}\right\}$$
(7.19)

where $x - h \leq x_3 \leq x + h$.

Equation (7.19) is a difference equation for $e_{j,m}$, the error in the solution of the finite difference approximation. We must examine whether this error grows as the solution proceeds. Convergence requires that $e_{j,m} \to 0$ as $h, k \to 0$ for all m.

Let E_m denote the largest in absolute value of the errors occurring at any time mk:

$$E_m = \max_{j} |e_{j,m}|$$

and let F be the absolute value of the maximum of this expression at any value of j and at any time from zero to mk:

$$F = \max_{j,m} \left|\left(\frac{\partial \Theta}{\partial t}\right)_{x,t_1} - \left(\frac{\partial^2 \Theta}{\partial x^2}\right)_{x_3,t}\right|$$

If we now impose the restriction $0 < r \leq \frac{1}{2}$, then $(1 - 2r) \geq 0$, so that, taking the absolute value of each term in (7.19) we obtain

$$|e_{j,m+1}| \leq (1 - 2r)|e_{j,m}| + r(|e_{j+1,m}| + |e_{j-1,m}|) + kF$$

Therefore

$$E_{m+1} \leq (1 - 2r)E_m + 2rE_m + kF$$
$$\leq E_m + kF$$

and therefore

$$E_m \leq E_{m-1} + kF$$

Applying this relationship repeatedly, we obtain

$$E_m \leq E_{m-1} + kF$$
$$\leq (E_{m-2} + kF) + kF$$
$$\leq E_{m-2} + 2kF$$
$$\vdots$$
$$\leq E_0 + mkF$$

However, $E_0 = 0$ because we would ensure that we have exact initial conditions, i.e. that $\theta(x, 0) = \Theta(x, 0)$. Since $mk = t$, we have that

$$E_m \leq tF$$

Now, as $h, k \to 0$,

$$F \to \left\{ \left(\frac{\partial \Theta}{\partial t} \right) - \left(\frac{\partial^2 \Theta}{\partial x^2} \right) \right\}_{x,t} = 0$$

because this expression is the original PDE. Therefore if $h, k \to 0$ in such a way that $0 < r \leq \frac{1}{2}$, and if t remains finite (i.e. we do not perform an infinite number of calculations) then $E_m \to 0$.

We have thus shown that the solution of the difference equation (7.10) converges to the solution of (7.4) if $0 < r \leq \frac{1}{2}$. This analysis has not considered what happens if $r > \frac{1}{2}$, although the example in Section 7.5 showed the process to be divergent for the particular value $r = 1$. In fact, it can be proven that (7.10) is divergent for $r > \frac{1}{2}$, and we say that the difference scheme is *conditionally convergent*, requiring $0 < r \leq \frac{1}{2}$. However, we shall not bother with the proof, since we shall shortly show that the simple explicit method is also *unstable* for $r > \frac{1}{2}$. Its convergence in this circumstance is irrelevant.

7.9 Stability

So far in this chapter it has been supposed that the calculations can be made exactly. We have neglected to consider the effect – the cumulative effect – of round-off errors which must almost always be present in any computations.

The effect of such errors in the use of the process defined by (7.10), for example, can be illustrated by artificially introducing an error at one point and computing the behaviour of θ at that and neighbouring points as time

Table 7.3 Error growth using the explicit method with $r = \frac{1}{2}$.

x		0	0.2	0.4	0.6	0.8	1.0
X (m)		0	0.1	0.2	0.3	0.4	0.5
τ (s)	t						
0	0.00	0	0	ε	0	0	0
400	0.02	0	$\varepsilon/2$	0	$\varepsilon/2$	0	0
800	0.04	0	0	$\varepsilon/2$	0	$\varepsilon/4$	0
1200	0.06	0	$\varepsilon/4$	0	$3\varepsilon/8$	0	0
1600	0.08	0	0	$5\varepsilon/16$	0	$3\varepsilon/16$	0
2000	0.10	0	$5\varepsilon/32$	0	$\varepsilon/4$	0	0

Table 7.4 Error growth using the explicit method with $r = 1$.

x		0	0.2	0.4	0.6	0.8	1.0
X (m)		0	0.1	0.2	0.3	0.4	0.5
τ (s)	t						
0	0.00	0	0	ε	0	0	0
400	0.02	0	ε	$-\varepsilon$	ε	0	0
800	0.04	0	-2ε	3ε	-2ε	ε	0
1200	0.06	0	5ε	-7ε	6ε	-3ε	0
1600	0.08	0	-12ε	18ε	-16ε	9ε	0
2000	0.10	0	30ε	-46ε	43ε	-25ε	0

passes. Since (7.4) is a linear equation, we can examine the error in θ independently of θ itself.

Let an error of ε occur at the point ($x = 0.4$, $t = 0$) of Table 7.1. The growth of the error, using (7.11) – or (7.19) without the truncation error term – is shown in Table 7.3.

It can be seen that the errors have spread outwards to affect all interior points. No errors occur at the boundaries $x = 0$ and 1, because θ is fixed at those points by the boundary conditions and is not calculated there. In the interior the errors oscillate in value, but with a decreasing amplitude. Eventually they will be found to decay to zero.

The use of (7.12), however, in which $r = 1$, produces a different result, as shown in Table 7.4. Now we see that the error increases at an increasing rate. Not only does Table 7.2 show that with $r = 1$ the explicit method is divergent*; Table 7.4 illustrates that the process is also unstable.

In general terms, we say that a difference scheme is stable if an error made at any stage of the calculation does not grow without limit as the calculation proceeds. Note that this definition allows a process to be termed stable if an error remains constant, or grows (with increasing time) at a decreasing rate, reaching some limit. Such a situation may be tolerable if the solution itself grows and remains large compared with the error.

One method for the study of the stability of a difference scheme involves the construction of a Fourier series solution of the difference equation. Consider, for example, the explicit formula (7.10), which we write as

$$\theta_{j,m+1} = (1 - 2r)\theta_{j,m} + r(\theta_{j+1,m} + \theta_{j-1,m}) \tag{7.20}$$

and suppose that we look for a solution of the form

$$\theta_{j,m} = \sum_n A_n(t) \sin nj\pi x \tag{7.21}$$

---

* It is emphasized that the wrong values in Table 7.2 are not due to instability, since the values there were calculated exactly. No round-off (or other) errors were made, and that Table does not illustrate the growth of any errors. The wrong values are due to lack of convergence.

or, equivalently,

$$\theta_{j,m} = \sum_n A_n(t)\, e^{inj\pi x}$$

where $i = \sqrt{(-1)}$ and, as usual, $0 \leq x \leq 1$.

Since the original differential equation (7.4) is linear, we need consider only one term of this series, viz.

$$\theta_{j,m} = A(t)\, e^{ij\pi x}$$

Since the solution of the conduction equation, with finite boundary conditions, must also remain finite as time increases, the coefficient $A(t)$ must remain bounded as $t \to \infty$. To ensure this, it is sufficient to put

$$A(t) = A^t = A^{mk} = B\omega^m \text{ say}$$

and require that

$$|\omega| \leq 1 \tag{7.22}$$

We shall examine the stability of (7.20) by using

$$\theta_{j,m} = B\omega^m\, e^{ij\pi x} \tag{7.23}$$

as a trial solution, and finding the value or values of ω which enable this expression to be a solution of (7.20). If ω then satisfies (7.22) we may conclude that (7.20) is a stable process.

Substitution of (7.23) into (7.20) leads to

$$B\omega^{m+1}\, e^{ij\pi x} = (1 - 2r)B\omega^m\, e^{ij\pi x} + rB\omega^m(e^{i(j+1)\pi x} + e^{i(j-1)\pi x})$$

or

$$\omega = (1 - 2r) + r(e^{i\pi x} + e^{-i\pi x})$$

Recalling that $e^{i\pi x} = \cos \pi x + i \sin \pi x$, we have that

$$\omega = (1 - 2r) + 2r \cos \pi x$$
$$= 1 - 4r \sin^2 \pi x/2 \tag{7.24}$$

The condition (7.22) then requires that

$$|1 - 4r \sin^2 \pi x/2| \leq 1 \tag{7.25}$$

Since $0 \leq \sin^2 \pi x/2 \leq 1$ for all values of x, (7.25) can only be satisfied if

$$1 - 4r \sin^2 \pi x/2 \leq 1 \quad \text{and} \quad -1 + 4r \sin^2 \pi x/2 \leq 1$$

i.e.

$$-4r \sin^2 \pi x/2 \leq 0 \quad \text{and} \quad 4r \sin^2 \pi x/2 \leq 2$$

i.e.

$$r \geq 0 \quad \text{and} \quad r \leq \tfrac{1}{2}$$

Clearly, the case $r = 0$ is not of interest, since this implies $\Delta t = 0$, under which condition the solution would not progress through time. It follows that (7.22) will be satisfied, and therefore (7.20) will be stable, if

$$0 < r \leq \tfrac{1}{2} \tag{7.26}$$

The practical consequence of this condition is that it places a fairly severe limit on the size of the time step that can be used. Recalling that $r = k/h^2$, and recognizing that to achieve any sort of accuracy we would require at least ten steps along the x-axis, we see that (7.26) imposes an upper value of 0.005 on the time step k. This means that 200 time steps must be completed for each unit of time – and it turns out that several units of (non-dimensional) time must be completed for most practical problems to reach steady state. If the explicit method is to be used for problems involving more complex calculations than those in the present simple example, then this can represent a substantial amount of computer time. Often, a method which permits the use of a larger time step is to be preferred. Such methods – in particular, implicit methods – are beyond the scope of this book.

It should be noticed that although the condition $0 < r \leq \frac{1}{2}$ is sufficient to ensure that $|\omega| \leq 1$, it allows ω to become negative. Since $\theta_{j,m} \propto \omega^m$, this means that the terms in (7.21) will oscillate in sign for some values of x, and therefore the solution – the sum of such terms – may also oscillate, or at least decay in an oscillating manner. For many problems, this is physically not correct and may therefore not be regarded as acceptable. To prevent it from happening, it is necessary to ensure that

$$0 \leq \omega \leq 1$$

which, from (7.24), requires

$$r \leq \tfrac{1}{4}$$

The increased severity of this condition means that, for many problems, the explicit method may not be suitable, despite its simplicity.

This method of examining the stability of a difference scheme is known as von Neumann stability analysis. It can generally only be applied to problems for which the difference schemes are linear and have constant coefficients. It also does not consider certain instabilities which can arise (in more difficult problems) from the boundary conditions. These limitations nevertheless allow the study of many important finite difference approximations; moreover, the requirement of constant coefficients can, to some extent, be relaxed (in a manner which is again beyond the scope of this book).

In Section 7.8 the comment was made that the difficulty in obtaining a proof of the convergence of an FDA was not serious. The reasons for this are that a stability analysis – at least the von Neumann stability analysis – is generally not too difficult, that consistency is also usually easy to demonstrate, and that there is a theorem known as Lax's equivalence theorem, which states that a consistent FDA to a properly posed, linear initial value PDE is convergent if and only if it is stable. Thus, for such a problem, if consistency and stability can be demonstrated, then convergence follows. The theorem applies only to properly posed* linear problems. Most

---

* In simple terms, a problem is properly posed if it has a bounded solution which is unique and which depends continuously on the data (i.e. a small change in one or other of the initial or boundary conditions causes only a small change in the solution).

(but certainly not all) problems arising in science and engineering which we may wish to solve by the methods discussed here are properly posed. However, many are not linear, so the theorem cannot be used. If convergence cannot be proven analytically, it can at least be demonstrated by numerical experiment. This should always be done, to validate numerical results.

7.10 An unstable finite difference approximation

As a further illustration of the von Neumann stability analysis, we will consider a method, known as Richardson's method, for the conduction equation.

A more accurate finite difference approximation than (7.10) uses a central difference approximation to the time derivative as well as to the space derivative:

$$\frac{\theta_{j,m+1} - \theta_{j,m-1}}{2k} = \frac{\theta_{j+1,m} - 2\theta_{j,m} + \theta_{j-1,m}}{h^2}$$

or

$$\theta_{j,m+1} = \theta_{j,m-1} + 2r(\theta_{j+1,m} - 2\theta_{j,m} + \theta_{j-1,m}) \qquad (7.27)$$

where $r = k/h^2$, as before.

Equation (7.27) is not self-starting: information (i.e. values of θ) at both $m = 0$ and $m = 1$ are needed before θ at $m = 2$ can be found. Nevertheless, it appears attractive because the central difference approximation to $\partial\theta/\partial t$ has a smaller truncation error than that in the FTCS method. From (7.17),

$$\frac{\theta_{j,m+1} - \theta_{j,m-1}}{2k} = \frac{\partial \theta}{\partial t} + \frac{k^2}{6} \frac{\partial^3 \theta}{\partial t^3} + O(k^4)$$

Thus, the truncation error of (7.27) is $O(k^2) + O(h^2)$, compared with $O(k) + O(h^2)$ for (7.10). It appears to be more accurate. Is it also stable?

As in the previous section, we assume a trial solution of the difference equation of the form

$$\theta_{j,m} = B\omega^m\, e^{ij\pi x} \qquad (7.23)$$

and determine the conditions, if any, necessary to ensure that $|\omega| \leq 1$. The substitution of (7.23) into (7.27) yields

$$B\omega^{m+1}\, e^{ij\pi x} = B\omega^{m-1}\, e^{ij\pi x} + 2r(e^{i(j+1)\pi x} - 2e^{ij\pi x} + e^{i(j-1)\pi x})B\omega^m$$

whence

$$\omega^2 = 1 - 8r\omega \sin^2 \pi x/2$$

This is a quadratic equation in ω with two roots which satisfy

$$\omega_1 \omega_2 = -1 \quad \text{and} \quad \omega_1 + \omega_2 = 8r \sin^2 \pi x/2 \qquad (7.28\text{a,b})$$

From (7.28a) if $|\omega_1| < 1$, then $|\omega_2| > 1$ and vice versa, and the method will be unstable. The only possibility is therefore $\omega_1 = -\omega_2 = 1$. But by (7.28b) this would cause $r = 0$ and therefore $k = 0$, which is not acceptable – if the time step k is zero, then the calculations will not make any progress. We therefore conclude that Richardson's method is unconditionally unstable.

7.11 Richardson's extrapolation

Richardson's method (7.27) for the solution of the conduction equation may not be of any use, but the extrapolation technique associated with his name, which we have already used several times, certainly is.

The FTCS method (7.10) has truncation errors which are $O(k) + O(h^2)$. We may therefore write the error – the difference between the analytical solution Θ and the numerical solution θ – as

$$\Theta - \theta_1 = Ck_1 + Dh_1^2 \qquad (7.29)$$

where C and D contain the derivatives $\partial^2 \theta/\partial t^2$ and $\partial^4 \theta/\partial x^4$ evaluated at appropriate (but unknown) values of x and t, and where k_1 and h_1 are one set of mesh sizes. If a second solution is found using k_2 and h_2, then assuming that C and D are constants,

$$\Theta - \theta_2 = Ck_2 + Dh_2^2 \qquad (7.30)$$

We have two equations, but three unknowns: Θ, C and D. We want to find Θ, and can therefore apparently eliminate either C or D, but not both.

If we eliminate D from (7.29) and (7.30), then we obtain

$$\Theta = \frac{h_2^2 \theta_1 - h_1^2 \theta_2}{h_2^2 - h_1^2} + \frac{k_1 h_2^2 - k_2 h_1^2}{h_2^2 - h_1^2} C \qquad (7.31)$$

Suppose now that the two space mesh sizes are such that $h_2 = h_1/2$, and that we impose an additional constraint: we keep $r = k/h^2$ constant. Then $k_2 = k_1/4$, and the coefficient of C in (7.31) vanishes. Θ is then given by

$$\Theta = \frac{h_2^2 \theta_1 - h_1^2 \theta_2}{h_2^2 - h_1^2} \qquad (7.32)$$

with an accuracy that is limited only by the quality of the assumption that C and D are constant. For this particular combination of values of h and k, the truncation error (7.32) will be $O(k^2) + O(h^4)$.

Worked examples

1. Solve the equation

$$\frac{\partial \theta}{\partial t} = \frac{\partial^2 \theta}{\partial x^2}$$

subject to the initial and boundary conditions

$$\theta(x, 0) = \sin \pi x \quad \text{for } 0 \leq x \leq 1, \quad t = 0$$
$$\theta(0, t) = 0 \quad \text{for } t > 0, \quad x = 0$$
$$\theta(1, t) = 0 \quad \text{for } t > 0, \quad x = 1$$

Use space step sizes (h) of 0.25, 0.125 and 0.0625 and time steps (k) chosen so that $k/h^2 = 0.5$. Apply Richardson's extrapolation to the solutions at $x = 0.25$ for the times $t = 0.125, 0.25$ and 0.5. In addition, compare (for the three step sizes) the dimensionless time and the number of time steps required until $\theta(0.5) \leq 0.01$.

(This is not a particularly realistic problem in itself, but it is important for two reasons: problems with more realistic initial or boundary conditions can be constructed from it by superposition, and it has a simple analytical solution

$$\Theta = \exp(-\pi^2 t) \sin \pi x$$

which can be used to determine the errors in the numerical solution.)

With $h = 0.25$ and $k = (0.25^2)/2 = 0.03125$, the first few results and the corresponding values of the error ($\Theta - \theta$) are as follows:

t			x		
	0.0	0.25	0.5	0.75	1.0
0.000 00	0.0000	0.7071	1.000	0.7071	0.0000
0.031 25	0.0000	0.5000	0.7071	0.5000	0.0000
errors	0.0000	0.1944E−01	0.2750E−01	0.1944E−01	0.0000
0.062 50	0.0000	0.3536	0.5000	0.3536	0.0000
errors	0.0000	0.2803E−01	0.3964E−01	0.2803E−01	0.0000
0.093 75	0.0000	0.2500	0.3536	0.2500	0.0000
errors	0.0000	0.3031E−01	0.4287E−01	0.3031E−01	0.0000
0.125 00	0.0000	0.1768	0.2500	0.1768	0.0000
errors	0.0000	0.2914E−01	0.4121E−01	0.2914E−01	0.0000
0.156 25	0.0000	0.1250	0.1768	0.1250	0.0000
errors	0.0000	0.2627E−01	0.3715E−01	0.2627E−01	0.0000
⋮	⋮	⋮	⋮	⋮	⋮

Notice that the solution is symmetrical about $x = 0.5$; by adopting a boundary condition $\partial\theta/\partial x = 0$ at $x = 0.5$ only half the solution domain need have been included in the solution process. An additional, hypothetical mesh point must be used, as described in Section 5.12.3.

With $h = 0.125$ and 0.0625, the corresponding values of k are 0.0078125 and 0.001953125, respectively. The solution is obtained in the same way as for $h = 0.25$, but with much greater effort. The solutions are not shown here, but students should obtain them for themselves.

The quantities sought in the question – the values of θ at the point $x = 0.25$ for the times $t = 0.125, 0.25$ and 0.5 – are given in the following table. θ_{12} and θ_{23} denote the extrapolated values obtained using h_1 and h_2, and then h_2 and h_3, respectively. Θ is the analytical solution. Adjacent to each θ, both the directly computed values and the extrapolated values, is given (in parentheses) the respective percentage error. The success of Richardson's extrapolation can be gauged from these error calculations.

	t		
	0.125	0.25	0.5
	θ (% error)	θ (% error)	θ (% error)
$h_1 = 0.25$	0.1768 (−14.1)	0.04419 (−26.3)	0.002762 (−45.7)
$h_2 = 0.125$	0.1992 (−3.3)	0.05613 (−6.4)	0.004455 (−12.4)
$h_3 = 0.0625$	0.2043 (−0.8)	0.05901 (−1.6)	0.004925 (−3.2)
θ_{12}	0.2067 (0.4)	0.06011 (0.2)	0.005019 (−1.3)
θ_{23}	0.2060 (0.04)	0.05997 (0.01)	0.005082 (−0.1)
Θ	0.2059	0.05997	0.005085

The full solution, not reprinted here, shows that with $k = 0.03125$ the mid-point temperature falls below 0.01 after 14 time steps, at $t = 0.4375$. With $k = 0.0078125$ the values are 59 and 0.46094; and with $k = 0.001953125$, they are 238 and 0.46484. According to the analytical solution, $\theta(0.5)$ falls to 0.01 at a time $t = 0.46660$.

2. Use the von Neumann method of stability analysis to investigate the use of the central difference approximation

$$\frac{\theta_{j,m+1} - 2\theta_{j,m} + \theta_{j,m-1}}{k^2} = \frac{\theta_{j+1,m} - 2\theta_{j,m} + \theta_{j-1,m}}{h^2}$$

to solve the *hyperbolic* wave equation $\partial^2\theta/\partial t^2 = \partial^2\theta/\partial x^2$, where $\theta(j, m)$ is defined as in Section 7.4.

We assume a trial solution of the form

$$\theta_{j,m} = B\omega^m e^{ij\pi x}$$

and obtain

$$B(\omega^{m+1} - 2\omega^m + \omega^{m-1})\,e^{ij\pi x}/k^2 = B(e^{i(j+1)\pi x} - 2e^{ij\pi x} + e^{i(j-1)\pi x})\omega^m/h^2$$

from which it follows that

$$\omega^2 - 2A\omega + 1 = 0$$

where $A = 1 - 2\lambda^2 \sin^2 \pi x/2$ and $\lambda = k/h$. The solutions of this quadratic are

$$\omega_1 = A + (A^2 - 1)^{1/2} \quad \text{and} \quad \omega_2 = A - (A^2 - 1)^{1/2}$$

By definition, $A \leq 1$. If $A < -1$, then $\omega_2 > 1$ and (7.22) shows that the process will be unstable. If $-1 \leq A \leq 1$, then $A^2 \leq 1$ and the values of ω can be written

$$\omega_1 = A + i(1 - A^2)^{1/2} \quad \text{and} \quad \omega_2 = A - i(1 - A^2)^{1/2}$$

Therefore $|\omega_1| = |\omega_2| = \{A^2 + (1 - A^2)\}^{1/2} = 1$ and (7.22) is satisfied. Stability therefore requires that A should lie between 1 and -1, and from the definition of A it follows that

$$-1 \leq 1 - 2\lambda^2 \sin^2 \pi x/2 \leq 1$$

The right-hand inequality is always satisfied; the left-hand inequality requires that $\lambda \leq 1$.

The final result is therefore that the process will be stable if $k \leq h$.

3. Use the FDA of the previous example to solve the wave equation subject to the boundary conditions

$$\theta = 0 \quad \text{at } x = 0 \quad \text{and} \quad x = 1$$

and the initial condition

$$\theta = \sin \pi x \quad \text{and} \quad \partial\theta/\partial t = 0 \quad \text{at} \quad t = 0$$

(Note that the the wave equation is second-order in time, and therefore requires *two* initial conditions.)

We will choose $\lambda = 1$, the limiting value for stability, and the FDA to the wave equation therefore becomes

$$\theta_{j,m+1} = \theta_{j+1,m} + \theta_{j-1,m} - \theta_{j,m-1} \tag{7.33}$$

This is a *three-level* scheme, meaning that values of θ at three values of time (i.e. at three values of m) appear: values of θ at time levels $m - 1$ and m are needed in order to compute θ at time $m + 1$. The first initial condition supplies values of θ at $m = 0$. The second initial condition can be replaced by the FDA

$$(\theta_{j,m+1} - \theta_{j,m-1})/k = 0$$

which, for $m = 0$, leads to

$$(\theta_{j,1} - \theta_{j,-1})/k = 0$$

or $\theta_{j,-1} = \theta_{j,1}$, where the subscript -1 denotes a fictitious time level at a time k units before $t = 0$. This enables us to apply the FDA to the wave equation at $m = 0$ to find values for θ at $t = k$, thereby obtaining the second set of values needed to continue the solution. The FDA for the particular case $m = 0$ becomes

$$\theta_{j,1} = \theta_{j+1,0} + \theta_{j-1,0} - \theta_{j,-1} = \theta_{j+1,0} + \theta_{j-1,0} - \theta_{j,1}$$

whence

$$\theta_{j,1} = \tfrac{1}{2}(\theta_{j+1,0} + \theta_{j-1,0}) \tag{7.34}$$

Equation (7.34) is used to start the solution, and (7.33) to continue it. Noting that the problem is symmetrical about $x = \tfrac{1}{2}$, we need compute only half the solution region. Using $h = k = 0.1$, the first few rows of the solution are:

				x			
	0	0.1	0.2	0.3	0.4	0.5	
t							
0.0	0.0000	0.3090	0.5878	0.8090	0.9511	1.0000	
0.1	0.0000	0.2939	0.5590	0.7694	0.9045	0.9511	
0.2	0.0000	0.2500	0.4755	0.6545	0.7694	0.8090	
0.3	0.0000	0.1816	0.3455	0.4755	0.5590	0.5878	
0.4	0.0000	0.0955	0.1816	0.2500	0.2939	0.3090	
0.5	0.0000	0.0000	0.0000	0.0000	0.0000	0.0000	
0.6	0.0000	-0.0955	-0.1816	-0.2500	-0.2939	-0.3090	

Problems

1. Solve the problem defined in the first worked example of Section 7.12 using the FTCS method, for a range of values of h and k satisfying $k/h^2 \leq \tfrac{1}{2}$, in order to develop a feel for the space and time step sizes needed to keep the error to a reasonable (say 0.1%) limit.

2. Solve the non-dimensional conduction equation (7.4) with the initial and boundary conditions

$$\theta(x, 0) = 0 \quad \text{for } 0 \leq x \leq 1, \quad t = 0$$
$$\theta(0, t) = 0 \quad \text{for } t > 0, \quad x = 0$$
$$\frac{\partial \theta}{\partial x}(1, t) = 1 \quad \text{for } t > 0, \quad x = 1$$

3. Solve the non-dimensional conduction equation (7.4) with the initial and boundary conditions

$$\theta(x, 0) = 0 \qquad \text{for } 0 \leq x \leq 1, \qquad t = 0$$
$$\theta(0, t) = \sin t \qquad \text{for } t > 0, \qquad x = 1$$
$$\theta(1, t) = 0 \qquad \text{for } t > 0, \qquad x = 0$$

Sketch the temperature–time history of the wall at various points across its thickness. How much non-dimensional time is required until the temperature variation at any point is periodic, i.e. until the initial condition is 'forgotten'?

Translate that into dimensional terms for a concrete wall which is (a) 0.5 m thick and (b) 0.05 m thick.

4. Consider the dimensional conduction equation (7.2) with the initial and boundary conditions

$$T(X, 0) = T_0 \qquad \text{for } 0 \leq X \leq L, \qquad \tau = 0$$
$$T(0, \tau) = T_0 + T_1 \sin \pi\tau/12 \qquad \text{for } \tau > 0, \qquad X = 0$$
$$\frac{\partial T}{\partial X}(L, \tau) = 0 \qquad \text{for } \tau > 0, \qquad x = L$$

in which the time (τ) is measured in hours. This equation describes the temperature distribution in a wall of thickness L, one face of which (at $X = L$) is insulated while the other (at $X = 0$) starts oscillating in temperature between $T_0 - T_1$ and $T_0 + T_1$ with a period of 24 h.

Choosing appropriate reference quantities for length, time and temperature, show that the equation can be made non-dimensional, but that the 'left-hand' boundary condition contains a parameter L^2/α.

Solve the problem for a range of values of this parameter.

Suppose that $T_0 = 300$ K and $T_1 = 285$ K. For what values of L will the temperature fluctuation at $X = L$ be no greater than 2 K, if the wall is made of

(a) wood ($\alpha = 1.02 \times 10^{-7}$ W m^{-1} K^{-1})
(b) brick ($\alpha = 7 \times 10^{-7}$ W m^{-1} K^{-1})
(c) steel ($\alpha = 1.2 \times 10^{-5}$ W m^{-1} K^{-1})?

5. Use the Dufort–Frankel method to solve the conduction equation subject to the conditions $\theta(x, 0) = \sin \pi x$, $\theta(0, t) = \theta(1, t) = 0$, using the following mesh sizes:

(a) $h = 0.1$, $k = 0.1$ $(k/h = 1)$
(b) $h = 0.025$, $k = 0.025$ $(k/h = 1)$
(c) $h = 0.01$, $k = 0.01$ $(k/h = 1)$
(d) $h = 0.1$, $k = 0.025$ $(k/h = 0.25)$
(e) $h = 0.1$, $k = 0.01$ $(k/h = 0.1)$

Compare the solutions with the analytical solution (see Worked Example 1 in Section 7.12), and with the FTCS solution which you should also obtain for a range of values of h and k.

8
Integral methods for the solution of boundary value problems

8.1 Introduction

In this book we have, so far, concentrated heavily on finite difference methods for the solution of boundary value problems. These methods were discussed in Sections 5.12 and 5.13 for ordinary differential equations and in Chapter 6 for elliptic partial differential equations.

There is another class of methods for the solution of such problems, known generally as integral methods. Included in this class are the *method of weighted residuals, finite element methods* and *boundary solution methods**.

A full, or even partial, discussion of these methods is beyond the scope of this book – they do not normally form part of a first course on numerical methods. This brief presentation is intended mainly to enable students to become aware that such methods exist. We shall restrict ourselves to a simple application to ordinary differential equations. We shall also show that each of these methods can be derived from a common basis and will therefore share the features demonstrated in earlier chapters using the finite difference method.

8.2 Integral methods

Integral approaches to the solution of differential equations can be demonstrated using a simple example. Consider the problem of determining a function $\phi(x)$ such that, between $x = 0$ and $x = L$

---

* Boundary solution methods are also known as both boundary integral methods and boundary element methods. We will use the name 'boundary solution'.

$$\frac{d^2\phi}{dx^2} - f = 0 \tag{8.1a}$$

with

$$\phi = 0 \quad \text{at} \quad x = 0 \text{ and } x = L \tag{8.1b}$$

where $f(x)$ is a known function of x. The basis of the methods lies in noting that the integral equation

$$\int_0^L w(x)\left(\frac{d^2\phi}{dx^2} - f\right) dx = 0 \tag{8.2}$$

will be satisfied by the solution of (8.1) for any function $w(x)$.

An approximate solution $\overline{\phi}(x)$ to the differential equation can be found by defining

$$\phi(x) \simeq \overline{\phi}(x) = \sum_{i=1}^n N_i(x) a_i \tag{8.3}$$

where each $N_i(x)$ is a function, known as a trial function, which we select; and each a_i is a coefficient which must be determined. For example, we could choose

$$N_1(x) = \sin \pi x/L \quad \text{and} \quad N_2(x) = \sin 2\pi x/L$$

Then (8.3) becomes

$$\overline{\phi}(x) = a_1 \sin \pi x/L + a_2 \sin 2\pi x/L.$$

We then obtain a system of n equations (in this case, $n = 2$) for the n unknown coefficients a_i in (8.3) by requiring that

$$\int_0^L w_i(x)\left(\frac{d^2\overline{\phi}}{dx^2} - f\right) dx = 0, \quad i = 1, 2, \ldots, n \tag{8.4}$$

where the $w_i(x)$ are n weighting functions which we also select and which control the sense in which the differential equation is satisfied.

For example, we could make $w_i(x)$ equal to $N_i(x)$:

$$w_1(x) = \sin \pi x/L \quad \text{and} \quad w_2(x) = \sin 2\pi x/L$$

As will be mentioned below, this leads to the weighted residual method. Alternatively, if $w_i(x)$ is chosen to be the Dirac function, defined by

$$\delta(x_i) = 0 \quad \text{for } x \neq x_i \quad \text{and} \quad \int_0^L \delta(x_i)\, dx = 1 \tag{8.5}$$

then the differential equation will be satisfied at the point $x = x_i$. The finite difference approach can be derived in this way. Clearly, care must be taken to choose suitable functions $N_i(x)$ and $w_i(x)$ so that the n equations in (8.4) can be solved for the n unknown coefficients a_i to define the approximate solution (8.3).

The introduction of the approximation (8.3) for $\phi(x)$ means that the equation will not, in general, be satisfied exactly for all points in the solution region. We can write

$$\frac{d^2\bar{\phi}}{dx^2} - f = R(x) \neq 0 \tag{8.6}$$

where $R(x)$ is the *residual*. Equation (8.4) becomes

$$\int_0^L w_i(x) R(x) \, dx = 0$$

that is, the residual is weighted by the function $w_i(x)$ and the integral forced to zero. For this reason solution procedures based on (8.4) are known as weighted residual methods.

To derive the finite element and boundary solution procedures, one further extension is required. We note that, using integration by parts,

$$\int_0^L w(x) \frac{d^2\phi}{dx^2} \, dx = -\int_0^L \frac{dw(x)}{dx} \frac{d\phi}{dx} \, dx + \left[w(x) \frac{d\phi}{dx} \right]_0^L$$

and

$$\int_0^L \frac{dw(x)}{dx} \frac{d\phi}{dx} \, dx = -\int_0^L \frac{d^2w(x)}{dx^2} \phi \, dx + \left[\frac{dw(x)}{dx} \phi \right]_0^L$$

It is therefore possible to write (8.2) in three equally valid forms:

$$\int_0^L w(x) \frac{d^2\phi}{dx^2} \, dx - \int_0^L w(x) f(x) \, dx = 0 \tag{8.7a}$$

$$-\int_0^L \frac{dw(x)}{dx} \frac{d\phi}{dx} \, dx + \left[w(x) \frac{d\phi}{dx} \right]_0^L - \int_0^L w(x) f(x) \, dx = 0 \tag{8.7b}$$

and

$$\int_0^L \frac{d^2w(x)}{dx^2} \phi \, dx - \left[\frac{dw(x)}{dx} \phi - w(x) \frac{d\phi}{dx} \right]_0^L - \int_0^L w(x) f(x) \, dx = 0 \tag{8.7c}$$

As we have seen, (8.7a) led to the method of weighted residuals.

The finite element method is usually based on (8.7b). The selection of the weighting functions $w_i(x)$ in (8.4) and the trial functions $N_i(x)$ in the approximation (8.3) is simplified by the fact that the highest order of differentiation within the integrals in (8.7b) is first order.

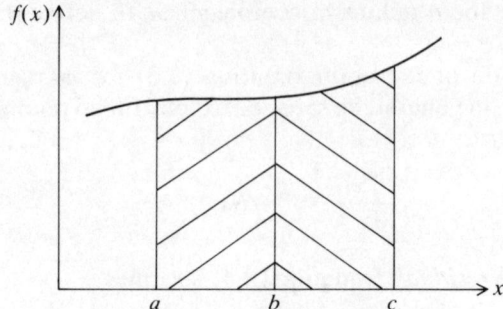

Figure 8.1 Integration over subdomains.

The finite element method also uses the property of integrals, shown in Figure 8.1, that if b lies between a and c, and if $f(x)$ is smooth, then

$$\int_a^c f(x)\,dx = \int_a^b f(x)\,dx + \int_b^c f(x)\,dx$$

The integral over the full solution region can be evaluated by summing the integrals over subregions, each of which becomes a 'finite element'.

The method known as the Galerkin finite element method is obtained if the weighting functions $w_i(x)$ are set equal to the trial functions $N_i(x)$ in (8.3). In general, the choice of weighting functions $w_i(x)$ and of approximation or trial functions $N_i(x)$ is limited only by the restriction that it must be possible to perform the integrals in (8.7b). A wide variety of finite element formulations is therefore possible.

A boundary solution method can be based on (8.7c). We choose $w(x)$ as the 'singular' solution to the governing differential equation, i.e. the solution when $f(x)$ is the Dirac function defined earlier. For the problem described by (8.1), $w(x)$ is the solution of

$$\frac{d^2w(x)}{dx^2} = \delta(x_i) \tag{8.8}$$

Then

$$\int_0^L \frac{d^2w(x)}{dx^2}\,\phi\,dx = \phi(x_i)$$

and (8.7c) becomes

$$\phi(x_i) = \left[\frac{dw}{dx}\phi - w\frac{d\phi}{dx}\right]_0^L + \int_0^L wf\,dx \tag{8.9}$$

We note that $\phi(x_i)$ is given in terms of the values of ϕ and $d\phi/dx$ at the ends of the solution region (the boundary points 0 and L) and an integral over the region of the known function $f(x)$.

8.3 Implementation of integral methods

In this section we will choose the approximations $N_i(x)$ and the weighting functions $w_i(x)$ to obtain, in turn, the central difference approximation, a weighted residual method, the linear Galerkin finite element approximation and the boundary solution method, for the problem (8.1). It will be seen that the derivation here of the finite difference method is not as simple as the approach taken in earlier chapters; however, it is worth including in order to illustrate the common basis for all these methods.

8.3.1 The central difference approximation

If we approximate $\phi(x)$ over the three points shown in Figure 8.2 by the quadratic function

$$\bar{\phi}(x) = p + qx + rx^2 \tag{8.10}$$

then

$$\bar{\phi}(0) \equiv \phi_1 = p$$
$$\bar{\phi}(h) \equiv \phi_2 = p + qh + rh^2$$
$$\bar{\phi}(2h) \equiv \phi_3 = p + 2qh + 4rh^2$$

Solving for p, q and r, we obtain

$$p = \phi_1$$
$$q = (-3\phi_1 + 4\phi_2 - \phi_3)/2h$$
$$r = (\phi_1 - 2\phi_2 + \phi_3)/2h^2$$

Let $\xi = x/2h$ and substitute for p, q and r into (8.10). Then

$$\begin{aligned}\bar{\phi}(x) &= \phi_1 + (-3\phi_1 + 4\phi_2 - \phi_3)\xi + (2\phi_1 - 4\phi_2 + 2\phi_3)\xi^2 \\ &= (1 - 3\xi + 2\xi^2)\phi_1 + (4\xi - 4\xi^2)\phi_2 + (-\xi + 2\xi^2)\phi_3 \\ &= N_1(x)\phi_1 + N_2(x)\phi_2 + N_3(x)\phi_3 \\ &= \sum_{i=1}^{3} N_i(x)\phi_i\end{aligned} \tag{8.11}$$

giving the functions $N_i(x)$ in (8.3).

Now selecting $w_i(x) = \delta(h)$, equation (8.4) with $L = 2h$ becomes

$$\int_0^{2h} \delta(h) \left(\frac{d^2\bar{\phi}}{dx^2} - f \right) dx = 0$$

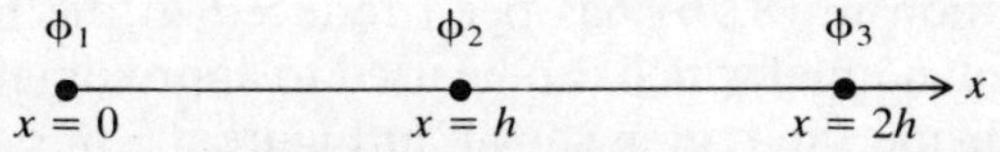

Figure 8.2 Grid points for the central difference approximation.

and using the properties of the Dirac function to delete the integral

$$\left[\frac{d^2\bar{\phi}}{dx^2} - f\right]_{x=h} = 0$$

Differentiating (8.11) twice and setting $x = h$ (i.e. $\xi = \frac{1}{2}$), we obtain

$$(\phi_1 - 2\phi_2 + \phi_3)/h^2 = f(h) \tag{8.12}$$

in which the left-hand side is the central difference approximation to $d^2\phi/dx^2$.

8.3.2 A weighted residual approximation

Let us now approximate the solution $\phi(x)$ by using, for (8.3),

$$\bar{\phi}(x) = a_1 \sin \pi x/L + a_2 \sin 2\pi x/L$$

We note that this choice of the approximating functions N_i immediately satisfies the boundary conditions $\phi(x) = 0$ at $x = 0$ and $x = L$. Then choosing the Galerkin weighting

$$w_1 \equiv N_1(x) = \sin \pi x/L \quad \text{and} \quad w_2 \equiv N_2(x) = \sin 2\pi x/L$$

(8.7a) leads to the two equations

$$-\frac{\pi^2}{L^2}\int_0^L \sin\frac{\pi x}{L}\left(a_1 \sin\frac{\pi x}{L} + 4a_2 \sin\frac{2\pi x}{L}\right) dx = \int_0^L \sin\frac{\pi x}{L} f(x)\, dx$$

and

$$-\frac{\pi^2}{L^2}\int_0^L \sin\frac{2\pi x}{L}\left(a_1 \sin\frac{\pi x}{L} + 4a_2 \sin\frac{2\pi x}{L}\right) dx = \int_0^L \sin\frac{2\pi x}{L} f(x)\, dx$$

Integrating and writing the equations in matrix form, we obtain

$$-\frac{\pi^2}{L^2}\begin{bmatrix} L/2 & 0 \\ 0 & 2L \end{bmatrix}\begin{Bmatrix} a_1 \\ a_2 \end{Bmatrix} = \begin{Bmatrix} \int_0^L \sin\frac{\pi x}{L} f(x)\, dx \\ \int_0^L \sin\frac{2\pi x}{L} f(x)\, dx \end{Bmatrix}$$

giving two equations which can be solved for a_1 and a_2.

8.3.3 The finite element method

The differentiation in (8.7b) has been reduced to the first order, so the lowest-order polynomial which can be used to approximate $\phi(x)$ is a linear function. We divide the region shown in Figure 8.3 into the two elements shown.

IMPLEMENTATION OF INTEGRAL METHODS

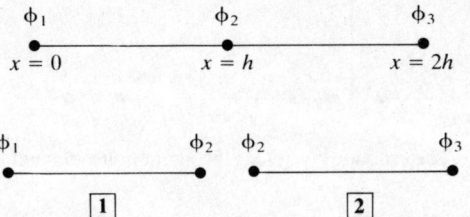

Figure 8.3 Subdivision of the solution domain into finite elements.

On element, $\boxed{1}$ we write the linear function

$$\bar{\phi}(x) = p_1 x + q_1 \tag{8.13}$$

and proceed to determine the coefficients p_1 and q_1 to satisfy the conditions $\bar{\phi} = \phi_1$ at $x = 0$ and $\bar{\phi} = \phi_2$ at $x = h$:

$$\bar{\phi}(0) \equiv \phi_1 = q_1 \qquad \bar{\phi}(h) \equiv \phi_2 = p_1 h + q_1$$

whence

$$p_1 = (\phi_2 - \phi_1)/h \qquad q_1 = \phi_1$$

Thus (8.13) becomes

$$\begin{aligned}\bar{\phi}(x) &= (\phi_2 - \phi_1)x/h + \phi_1 \\ &= (1 - x/h)\phi_1 + (x/h)\phi_2 \\ &= N_1(x)\phi_1 + N_2(x)\phi_2\end{aligned} \tag{8.14}$$

Similarly, for element $\boxed{2}$

$$\bar{\phi}(x) = p_2 x + q_2$$

with

$$\bar{\phi}(h) \equiv \phi_2 = p_2 h + q_2$$

and

$$\bar{\phi}(2h) \equiv \phi_3 = p_2(2h) + q_2$$

After we solve for p_2 and q_2, $\bar{\phi}(x)$ on element $\boxed{2}$ becomes

$$\begin{aligned}\bar{\phi}(x) &= (\phi_3 - \phi_2)x/h + 2\phi_2 - \phi_3 \\ &= (2 - x/h)\phi_2 + (x/h - 1)\phi_3 \\ &= N_2(x)\phi_2 + N_3(x)\phi_3\end{aligned} \tag{8.15}$$

The plot of the functions $N_i(x)$ in Figure 8.4 displays what are known as the 'hat functions' of the linear finite element approximation.

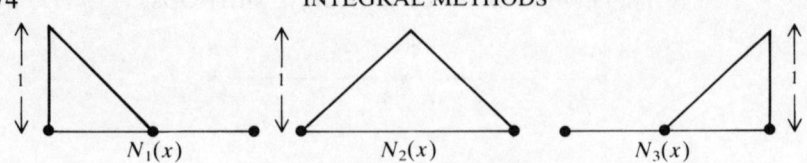

Figure 8.4 The functions $N_i(x)$ for the linear finite element approximation.

If we select the Galerkin weighting $w_i(x) = N_i(x)$, then (8.7b) will yield three equations for the three unknown quantities ϕ_1, ϕ_2 and ϕ_3. These particular weighting functions, now also shown in Figure 8.4, have a similar local character to the Dirac function used to derive the central difference approximation, but extend over the whole of the element or elements attached to each node, rather than acting only at a single point.

For example, the second of the three equations derived from (8.7b) is obtained by setting $w(x) = N_2(x)$. Then (8.7b) becomes

$$-\int_0^{2h} \frac{dN_2(x)}{dx} \frac{d\bar{\phi}}{dx}\, dx + \left[N_2(x) \frac{d\bar{\phi}}{dx}\right]_0^{2h} = \int_0^{2h} N_2(x) f(x)\, dx \quad (8.16)$$

We note that $N_2(x) = 0$ at $x = 0$ and at $x = 2h$; the second term on the left therefore vanishes. Then performing the integrals over each element separately, and substituting for $\bar{\phi}(x)$ from (8.14) and (8.15), we obtain

$$-\int_0^h \left(\frac{1}{h}\right)\left(-\frac{1}{h}\phi_1 + \frac{1}{h}\phi_2\right) dx - \int_h^{2h} \left(-\frac{1}{h}\right)\left(-\frac{1}{h}\phi_2 + \frac{1}{h}\phi_3\right) dx$$

$$= \int_0^h \left(\frac{x}{h}\right) f(x)\, dx + \int_h^{2h} \left(2 - \frac{x}{h}\right) f(x)\, dx$$

whence

$$\phi_1 - 2\phi_2 + \phi_3 = h\int_0^{2h} N_2(x) f(x)\, dx \quad (8.17)$$

Comparing (8.12) and (8.17), we see that this linear finite element approximation generates the same function of ϕ on the left-hand side as the central difference operator, and will give an identical solution for ϕ_i if $f(x)$ is constant. The right-hand side is different if, as would be the case in practice, $f(x)$ is not constant. The left-hand side also becomes different if the elements are not of equal length.

The character of the finite element method becomes more apparent if we complete the integrals in all of the equations over each element separately. Students should verify that on element $\boxed{1}$

$$\begin{bmatrix} \frac{1}{h} & -\frac{1}{h} & 0 \\ -\frac{1}{h} & \frac{1}{h} & 0 \\ 0 & 0 & 0 \end{bmatrix} \begin{Bmatrix} \phi_1 \\ \phi_2 \\ \phi_3 \end{Bmatrix} = \begin{Bmatrix} F_1^1 \\ F_2^1 \\ 0 \end{Bmatrix}$$

and on element $\boxed{2}$

$$\begin{bmatrix} 0 & 0 & 0 \\ 0 & \dfrac{1}{h} & -\dfrac{1}{h} \\ 0 & -\dfrac{1}{h} & \dfrac{1}{h} \end{bmatrix} \begin{Bmatrix} \phi_1 \\ \phi_2 \\ \phi_3 \end{Bmatrix} = \begin{Bmatrix} 0 \\ F_2^2 \\ F_3^2 \end{Bmatrix}$$

where h is the length of the element and

$$F_l^n = \int_{\text{element } n} N_l(x)\, f(x)\, \mathrm{d}x$$

with the subscript referring to node (mesh point) l and the superscript referring to element n over which the integral is evaluated.

In general, the coefficients in the matrix for element m have the form

$$k_{ij} = \int_{\text{element } m} \frac{\mathrm{d}N_i}{\mathrm{d}x} \frac{\mathrm{d}N_j}{\mathrm{d}x}\, \mathrm{d}x$$

Finally, the 'assembly' procedure to add the contributions from all elements corresponds to the summation of the integrals depicted in Figure 8.1. Adding the above matrices, we obtain the final equations in matrix form:

$$\begin{bmatrix} \dfrac{1}{h} & -\dfrac{1}{h} & 0 \\ -\dfrac{1}{h} & \dfrac{2}{h} & -\dfrac{1}{h} \\ 0 & -\dfrac{1}{h} & \dfrac{1}{h} \end{bmatrix} \begin{Bmatrix} \phi_1 \\ \phi_2 \\ \phi_3 \end{Bmatrix} = \begin{Bmatrix} F_1^1 \\ F_2^1 + F_2^2 \\ F_3^2 \end{Bmatrix}$$

8.3.4 A boundary solution procedure

A boundary solution procedure is obtained from (8.7c) if $w(x)$ satisfies (8.8). In other words, $w(x)$ must be the fundamental singular solution of the relevant governing differential equation. In the case of (8.8), that solution is

$$w(x) = -0.5(L - |r|) \qquad (8.18)$$

where r is the distance $x - x_i$. The value of $w(x)$ depends on whether x is less than or greater than x_i. We must emphasize this by the notation $w(x, x_i)$, defined as

$$w(x, x_i) = -0.5(L - (x_i - x)) \qquad \text{for } x < x_i \qquad (8.19a)$$
$$w(x, x_i) = -0.5(L - (x - x_i)) \qquad \text{for } x > x_i \qquad (8.19b)$$

To check that $w(x, x_i)$ defined in this way is the solution of (8.8), recall the properties of the Dirac function as stated in (8.5). Clearly, $d^2w(x, x_i)/dx^2 = 0$ for all points except $x = x_i$. In addition, for an arbitrarily small distance ε,

$$\int_0^L \frac{d^2w(x, x_i)}{dx^2} \, dx = \int_{x_i-\varepsilon}^{x_i+\varepsilon} \frac{d^2w(x, x_i)}{dx^2} \, dx$$

$$= \frac{dw(x, x_i)}{dx}\bigg|_{x_i+\varepsilon} - \frac{dw(x, x_i)}{dx}\bigg|_{x_i-\varepsilon}$$

$$= 1$$

using (8.18). This confirms that $w(x, x_i)$ is the required fundamental solution of the differential equation.

We substitute $w(x, x_i)$ for $w(x)$ in (8.7c), obtaining

$$\phi(x_i) = \int_0^L w(x, x_i) f(x) \, dx - \left[w(x, x_i) \frac{d\phi}{dx} - \frac{dw(x, x_i)}{dx} \phi \right]_0^L \quad (8.20)$$

We now consider two points x_i in the solution region which are within a small distance ε of the boundary points $x = 0$ and $x = L$. For these two points, (8.20) may be written in the form

$$\begin{Bmatrix} \phi(0+\varepsilon) \\ \phi(L-\varepsilon) \end{Bmatrix} = - \begin{bmatrix} \dfrac{dw(x, 0+\varepsilon)}{dx}\bigg|_0 & -\dfrac{dw(x, 0+\varepsilon)}{dx}\bigg|_L \\ \dfrac{dw(x, L-\varepsilon)}{dx}\bigg|_0 & -\dfrac{dw(x, L-\varepsilon)}{dx}\bigg|_L \end{bmatrix} \begin{Bmatrix} \phi(0) \\ \phi(L) \end{Bmatrix}$$

$$+ \begin{bmatrix} w(x, 0+\varepsilon)|_0 & -w(x, 0+\varepsilon)|_L \\ w(x, L-\varepsilon)|_0 & -w(x, L-\varepsilon)|_L \end{bmatrix} \begin{Bmatrix} \dfrac{d\phi}{dx}\bigg|_0 \\ \dfrac{d\phi}{dx}\bigg|_L \end{Bmatrix}$$

$$+ \begin{Bmatrix} \int_0^L w(x, 0+\varepsilon) f(x) \, dx \\ \int_0^L w(x, L-\varepsilon) f(x) \, dx \end{Bmatrix}$$

and, taking the limit as $\varepsilon \to 0$

$$\begin{Bmatrix} \phi(0) \\ \phi(L) \end{Bmatrix} = \begin{bmatrix} 0.5 & 0.5 \\ 0.5 & 0.5 \end{bmatrix} \begin{Bmatrix} \phi(0) \\ \phi(L) \end{Bmatrix} - \begin{bmatrix} 0.5L & 0 \\ 0 & -0.5L \end{bmatrix} \begin{Bmatrix} \dfrac{d\phi(0)}{dx} \\ \dfrac{d\phi(L)}{dx} \end{Bmatrix}$$

$$+ \begin{Bmatrix} \int_0^L w(x, 0) f(x) \, dx \\ \int_0^L w(x, L) f(x) \, dx \end{Bmatrix} \quad (8.21)$$

whence

$$\begin{bmatrix} 0.5 & -0.5 \\ -0.5 & 0.5 \end{bmatrix} \begin{Bmatrix} \phi(0) \\ \phi(L) \end{Bmatrix} + \begin{bmatrix} 0.5L & 0 \\ 0 & -0.5L \end{bmatrix} \begin{Bmatrix} \dfrac{d\phi(0)}{dx} \\ \dfrac{d\phi(L)}{dx} \end{Bmatrix}$$

$$= \begin{Bmatrix} \int_0^L w(x,0)\, f(x)\, dx \\ \int_0^L w(x,L)\, f(x)\, dx \end{Bmatrix} \quad (8.22)$$

The system (8.22) represents two equations between four quantities: $\phi(0)$, $\phi(L)$, $d\phi(0)/dx$ and $d\phi(L)/dx$. We note that the original problem (8.1) can be modified slightly: values of either ϕ or $d\phi/dx$ could be given as boundary conditions at $x = 0$ and $x = L$ (although if the derivative is specified at both boundaries the problem can only be solved to within an arbitrary constant). These two equations may be used to obtain the two unknown values of either ϕ or $d\phi/dx$, whichever is not the boundary condition, at each end of the solution region.

Equation (8.20) can then be used to evaluate $\phi(\xi)$ at any interval point ξ. The values of $d\phi(\xi)/d\xi$ can also be found directly. From (8.20),

$$\phi(\xi) = \int_0^L w(x,\xi)\, f(x)\, dx - \left[w(x,\xi)\frac{d\phi}{dx} - \frac{dw(x,\xi)}{dx}\phi \right]_{x=0}^{x=L}$$

and, differentiating,

$$\frac{d\phi}{d\xi} = \int_0^L \frac{dw(x,\xi)}{d\xi} f(x)\, dx - \left[\frac{dw(x,\xi)}{d\xi}\frac{d\phi}{dx} - \frac{d^2w(x,\xi)}{d\xi dx}\phi \right]_{x=0}^{x=L}$$
(8.23)

It is worth noting here that it was not necessary, in this one-dimensional example, to define the approximate solution (8.3). However, this approximation is an essential step in two- and three-dimensional formulations of the boundary solution procedure where the integration by parts leading to (8.7c) gives the equation

$$\phi(x_i) = \int_S \phi \frac{dw(x)}{dx}\, dS(x) - \int_S w(x)\frac{d\phi}{dx}\, dS(x) + \int_V w(x) f(x)\, dV(x)$$
(8.24)

where S is the surface of the solution region and V is its interior. In order to set up a solution procedure, $\phi(x)$ and $d\phi/dx$ must be defined in terms of discrete values using an approximation in the form of (8.3). Both $\phi(x)$ and $d\phi/dx$ in (8.24) are thus given in terms of 'nodal' quantities ϕ_i and $d\phi/dx_i$, and a known variation $N_i(x)$ between nodes which can be integrated. Sufficient points x_i are chosen on the boundary to give a set of equations similar to (8.22), which must be solved for the unknown boundary values. Equation (8.24) can then be used to evaluate ϕ at any required internal point. If, as is often the case, $f(x) = 0$, then the division into 'elements' is confined to the surface. It is this feature which has led to the name 'boundary element method'.

Worked examples

We will demonstrate the application of the finite element and boundary solution methods to the solution of

$$\frac{d^2\phi}{dx^2} = -x^2 \qquad (8.25)$$

with $\phi(0) = \phi(1) = 0$. The exact solution of this problem is

$$\phi(x) = x(1 - x^3)/12$$

The finite element method

For the implementation of the finite element method, we will choose to divide the solution region into the elements shown in Figure 8.5a. This mesh has the desirable feature that more elements are located at the right-hand end of the solution region where, because $d^2\phi/dx^2$ is largest there, we expect the greatest variation in $d\phi/dx$. Using a linear element like (8.16), $d\phi/dx$ is constant along the element, and it is therefore desirable to use smaller elements in such regions.

The evaluation of the integrals in (8.7b) is made easier if we recognize that the functions N_i appearing in the approximation for ϕ are non-zero only on those elements connected to node i, and take the form shown in Figure 8.5b. Associated with node i are elements $\boxed{i-1}$ and $\boxed{i}$ (except for the first and last nodes, which are connected to only one element). If we introduce a co-ordinate ζ within each element (running from left to right), and if the length of element $\boxed{i}$ is h_i, then for an element $\boxed{i}$,

$$N_{i-1}(\zeta) = 1 - \zeta/h_i \quad \text{and} \quad N_i(\zeta) = \zeta/h_i$$

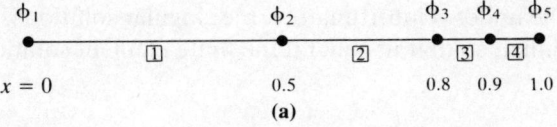

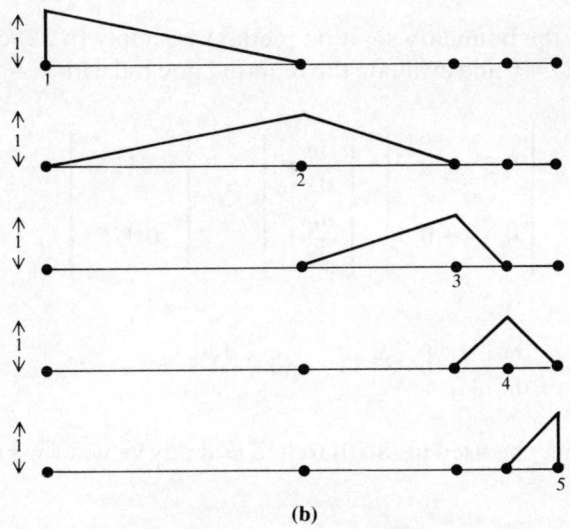

Figure 8.5 The finite element mesh and the functions $N_i(x)$ for (8.25).

Evaluating the integrals in (8.7b), we can write the resulting equations in matrix form

$$\begin{bmatrix} 2 & -2 & 0 & 0 & 0 \\ -2 & 5.3333 & -3.3333 & 0 & 0 \\ 0 & -3.3333 & 13.3333 & -10 & 0 \\ 0 & 0 & -10 & 20 & -10 \\ 0 & 0 & 0 & -10 & 10 \end{bmatrix} \begin{Bmatrix} 0 \\ \phi_2 \\ \phi_3 \\ \phi_4 \\ 0 \end{Bmatrix} = \begin{Bmatrix} 0.010\,42 - \dfrac{d\phi}{dx}\bigg|_0 \\ 0.086\,00 \\ 0.109\,00 \\ 0.081\,17 \\ 0.046\,75 + \dfrac{d\phi}{dx}\bigg|_L \end{Bmatrix}$$

giving the solution

$$\phi_2 = 0.036\,46 \qquad \phi_3 = 0.032\,53 \qquad \phi_4 = 0.020\,33$$

$$\frac{d\phi}{dx}\bigg|_0 = 0.083\,33 \qquad \frac{d\phi}{dx}\bigg|_L = -0.25$$

The nodal values of ϕ and $d\phi/dx$ are exact. This occurs when the approximation (8.3) can represent exactly the singular solution (8.19) when

The boundary solution method

To implement the boundary solution method we apply (8.22) directly. We set $\phi(0) = \phi(1) = 0$ and evaluate the required integral with $L = 1$. Equation (8.22) yields

$$\begin{bmatrix} 0.5 & 0 \\ 0 & -0.5 \end{bmatrix} \left\{ \begin{array}{c} \dfrac{d\phi}{dx}\bigg|_0 \\ \dfrac{d\phi}{dx}\bigg|_1 \end{array} \right\} = \left\{ \begin{array}{c} 0.04167 \\ 0.125 \end{array} \right\}$$

Therefore

$$\frac{d\phi}{dx}\bigg|_0 = 0.08333 \quad \text{and} \quad \frac{d\phi}{dx}\bigg|_1 = -0.25$$

These values may be used in (8.20) to find ϕ at any value of x. For example, at $x = 0.5$,

$$\phi(0.5) = -\left[w(x, 0.5) \frac{d\phi}{dx} - \frac{dw(x, 0.5)}{dx} \phi \right]_0^1 + \int_0^1 w(x, 0.5) f(x) \, dx$$

with

$$w(x, 0.5) = \begin{cases} -0.5(0.5 + x) & \text{for } x < 0.5 \\ -0.5(1.5 - x) & \text{for } x > 0.5 \end{cases}$$

Then

$$\phi(0.5) = [(0.5)(1.5 - 1)(-0.25) - (0.5)(0.5)(0.08333)]$$
$$- \int_0^{0.5} (0.5)(0.5 + x)(-x^2) \, dx - \int_{0.5}^1 (0.5)(1.5 - x)(-x^2) \, dx$$
$$= -0.08333 + 0.11979$$
$$= 0.03646$$

We note that no approximation has been required in this one-dimensional implementation of the boundary solution method. The solution is therefore exact. For two- and three-dimensional problems, the use of (8.3) to define an approximation on the surface in terms of discrete values of ϕ_i and $d\phi/dx_i$ will lead to a solution which generally will not be exact.

Problems

1. Solve the worked example (8.25) using (a) the finite element method and (b) the boundary solution method, but with the boundary conditions

$$\phi(0) = 0 \qquad \frac{d\phi}{dx}(1) = 0$$

The exact solution is

$$\frac{x}{12}(4 - x^3)$$

2. Solve the equation

$$\frac{d^2\phi}{dx^2} = x^3 \qquad \text{with } \phi(0) = \phi(1) = 0$$

by the method of weighted residuals. Assume that

$$\overline{\phi}(x) = \sum_{i=1}^{n} a_i \sin \frac{i\pi x}{L}$$

Solve the problem for $n = 1, 2$ and 3. Use the Galerkin weighting, and compare your three solutions at $x = 0.25$

[Hint: You may wish to evaluate the integrals $\int_0^1 x^3 \sin i\pi x \, dx$ numerically. Use any of the methods of Chapter 4. Perform the integration to sufficient accuracy to ensure that the integrating error is negligible.]

3. Solve the equation

$$y'' - y' + x = 0 \qquad \text{with } y(0) = y(1) = 0$$

using two equal length linear finite elements.
[Hint: Equation (8.17) becomes

$$-\int_0^{2h} \frac{dN_2}{dx} \frac{dy}{dx} dx + \left[N_2 \frac{dy}{dx} \right]_0^{2h} - \int_0^{2h} N_2 y' \, dx = -\int_0^{2h} N_2 x' \, dx$$

4. Solve the equation

$$\frac{d^2\phi}{dx^2} = 1 \qquad \text{with } \phi(0) = \phi(1) = 0$$

using three equal length linear finite elements. Draw a graph of the finite element solution $\overline{\phi}(x)$; note that it will be piecewise linear – being exact at the nodes, but not between them. Find the errors in $d\phi/dx$ at each end of each element, and at the mid-point of each element.

Repeat the solution using six elements, and compare the errors.

5. Solve the equation

$$\frac{d^2\phi}{dx^2} = x^3$$

subject to the boundary conditions

$$\phi = 0 \text{ at } x = 0 \quad \text{and} \quad d\phi/dx = 1 \text{ at } x = 1$$

by the boundary element method. Find ϕ and $d\phi/dx$ at $x = 0.5$, and compare these results with the exact values.

Suggestions for further reading

For those who wish to go further into these topics than this introductory text has taken them, or who seek a firmer theoretical foundation, the following books are recommended:

Bannerjee, P. K. and R. Butterfield, 1981. *Boundary element methods in engineering science*. London: McGraw-Hill.
Carnahan, B., H. A. Luther & J. O. Wilkes 1969. *Applied numerical methods*. New York: Wiley.
Fried, I. 1979. *Numerical solution of differential equations*. New York: Academic Press.
Hildebrand, F. B. 1974. *Introduction to numerical analysis*, 2nd edn. New York: McGraw-Hill.
Isaacson, E. & H. B. Keller 1966. *Analysis of numerical methods*. New York: Wiley.
Mitchell, A. R. & D. F. Griffiths 1980. *The finite difference method in partial differential equations*. New York: Wiley.
Smith, G. D. 1978. *Numerical solution of partial differential equations*, 2nd edn. Oxford: Clarendon Press.
Zienkiewicz, O. C. and K. Morgan 1983. *Finite elements and approximation*. New York: Wiley.

Index

Adams' method for O.D.E. 171
algebraic equations 14
algorithm 8
average operator 118

backward difference approximation
  differentiation 132
  integration 142
  interpolation 130
backward difference operator 118
backward difference table 124–5
Bairstow's method 56–8
biharmonic equation 238–9
block tridiagonal systems 87–8
boundary conditions 180
  elliptic equations 214, 223–5
  boundary value problems 186–8
boundary solution method 270, 275–8
boundary value problems (O.D.E.) 186–99
  derivative boundary conditions 193–4
  finite difference method 187–92
  non-linear equations 198–9
  Richardson's extrapolation 192–3
  shooting method 195–7

central difference approximation
  boundary value problems 187–92
  differentiation 133–4
  integral methods 271–2
  interpolation 131
central difference table 125
central difference operator 118
conduction equation 210, 214, 243–4
consistency 221
  Dufort–Frankel method 252–3
  FTCS method 252
  Lax's theorem 259
  Richardson's method 260–1
convergence
  damped iteration 35–6
  FTCS method 254–6
  iterative 25, 89, 222
  Lax's theorem 259
  mesh size 222, 253–6
  Newton's method 37
  simple iteration 26–31
convergence criterion
  Gauss–Seidel iteration 91
  Jacobi iteration 90

Newton's method, non-linear system 106–8
Newton's method, single equation 38, 39
simple iteration, non-linear system 104–5
simple iteration, single equation 26

Descartes' rule of signs 47
diagonal dominance 90
difference tables
  backward 124–5
  central 125
  forward 123–4
differential correction 100–3
differentiation, numerical *see* numerical differentiation
diffusion equation 210
Dirac function 268
direct method 2–3
Dufort–Frankel method 252–3
  consistency 252–3

elimination 73–94
elliptic equations
  biharmonic equation 238–9
  boundary conditions 214
  central difference approximation 213–14
  definition 210
  non-rectangular regions 231–7
  solution mesh 212
errors 8–11
Euler's method for O.D.E. 158–63
  modified Euler method 165–8
    convergence criterion 167–8
existence of solutions 3
extrapolation
  Aitken's 32–4
  Richardson's 146–8, 192–3, 222–3, 261

finite difference operators 117, 118–22
  average 118
  backward 118
  central 118
  forward 118
  identity 118
  integral 119
  shift 118
finite element method 269, 272–5
  Galerkin method 270, 274
first-order process 26

INDEX

forward difference approximation
  differentiation 132
  integration 141
  interpolation 127
foward difference operator 119
forward difference table 123–4
forward time, central space method (FTCS) 246–51
  consistency 251–2
  stability 256–60

Gaussian elimination 75–8
Gauss–Seidel iteration 91–2, 198, 217
global truncation error 163

Hamming's method for O.D.E. 172
hyperbolic equations
  definition 211
  stability analysis of central difference approximation 263–4

identity operator 120
initial estimate 15–19
  elliptic equation 217
  Newton's method 37
integral operator 119
integration, numerical *see* numerical integration
interpolation 125–32
  accuracy 128–9
  error reduction 147
  linear 126
  Newton's backward formula 130
  Newton's forward formula 126–30
  polynomial 126
    choice of 130
  quadratic 127
  Stirling's formula 131
interval halving 19–24
inversion 94–6
iterations, termination of 31, 41–2
iterative method 2, 14
iterative methods for single equations
  damped simple iteration 34–7
  interval halving 19–24
  Newton's method 37–45
  polynomial equations 47–58
    Bairstow's method 56–8
  regula falsi 46
  secant method 46
  simple iteration 24–5
iterative methods for systems of equations
  elliptic equations 217–21
  Gauss–Seidel iteration 91–2, 198, 217
  Jacobi iteration 89–91, 198
  modified Newton's method for non-linear systems 107–8

Newton's method for non-linear systems 106–8
successive over-relaxation 92–4
successive under relaxation 94

Jacobi iteration 89–91, 198

Laplace's equation 210
Lax's theorem 259
least squares 96–9
linear systems
  block tridiagonal 87–8
  elimination 73–94
  Gaussian elimination 75–8
  Gauss–Seidel iteration 91–2
  Jacobi iteration 89–91
  pentadiagonal 86–7
  successive over-relaxation (SOR) 92–4
  successive under-relaxation (SUR) 94
  tridiagonal 83–5
local truncation error 161

matrix inversion 94–6
Milne's method 170–2, 173, 174
  stability 175

Navier–Stokes equations 211
Newton's interpolation formula
  backward 130
  forward 126–30
Newton's method (Newton–Raphson) 37–43
  extended 43–6
  non-linear systems 106–8
Newton's relations 48–9
non-dimensional equations 11–12, 215–16, 244–5
non-linear boundary value problems 198–9
  convergence 199
  shooting method 199
non-linear systems
  modified Newton's method 107–8
  Newton's method 106–8
  simple iteration 103–5
numerical analysis 3
numerical differentiation 132–4
  backward formula 132
  central formula 133
  error reduction 147
  forward formula 132
  higher order 134
  non-tabular points 138–9
  summary of formulae Tables 4.6–8
numerical integration 139–46
  backward formula 142
  error reduction 148
  forward formula 141
  Simpson's rule 142, 143–4
  summary of formulae 143–6

INDEX

numerical integration (*contd*)
  trapezoidal rule 142, 143
numerical method 2

ordinary differential equations (O.D.E.)
  Adams' method 171
  boundary value problems 186–99
  Euler's method 158–63
  Hamming's method 172
  higher-order equations 180–2
  Milne's method 170–1, 173, 174
  modified Euler method 165–8
  predictor–corrector methods 168–76
  Runge–Kutta methods 176–9
  systems of equations 180–2
  Taylor series method 163–5
over-relaxation 92–4

parabolic equations
  conduction equation 243–4
  consistency 251–2
  definition 210
  Dufort–Frankel method 252–3
  explicit method (FTCS) 246–52
pentadiagonal systems 86–7
Poisson's equation 210
polynomial equations 47–58
  Bairstow's method 56–8
  Descartes' rule of signs 47
  Newton's relations 47–9
  number of roots 47
  synthetic division 50–3
predictor–corrector (P–C) methods 168–76
  Adams' methods 171
  error estimation 174–6
  general corrector formula 170
  general predictor formula 169
  Hamming's method 172
  Milne's method 170–2, 173, 174
  modified Euler method 165–8
  starting procedure 172–4

regula falsi 46
relaxation
  manual 225–31
  SOR 92–4, 217
  SUR 94
Richardson's extrapolation
  boundary value problems 192–3
  differentiation 147–8
  elliptic equations 222–3
  integration 148
  interpolation 147
  parabolic equations 261
Richardson's method 260–1
round-off errors 8–9, 10–11, 34
Runge–Kutta methods 176–9
  error estimation 179

higher-order equations 180
Runge–Kutta–Merson method 179
step size control 179
systems of equations 180–2

secant method 46–7
second-order process 37
shift operator 118
shooting method 195–8
simple iteration 24–31
  damped 35–6
Simpson's rule 142, 143–4
stability
  FTCS method 256–60
  Lax's theorem 259
  Milne's method 175
  Richardson's method 260–1
  von Neumann's method 257–60
step size adjustment
  predictor–corrector methods 175–6
  Runge–Kutta–Merson method 179
successive over-relaxation (SOR) 92–4, 217
  optimum SOR factor for Poisson's equation 218
synthetic division 50–3
systems of equations
  see linear systems
  non-linear systems

Taylor series method for O.D.E. 163–5
termination of iterations 31, 41–2
Thomas algorithm 83–5
transcendental equations 14
trapezoidal rule 142, 143
tridiagonal systems 83–5
truncation error 9–11
  Adams' method 171
  differentiation formulae 134–6
  Euler's method 161
  Hamming's method 172
  integration formulae 142
  Milne's method 170, 172–3
  modified Euler method 172–3
  predictor–corrector methods 174–5
  reduction of 146–8, 192–3, 222–3, 261
  Runge–Kutta–Merson method 179
  Runge–Kutta method 178

under-relaxation 94
uniqueness of solutions 3

von Neumann stability analysis
  FTCS method 257–60
  wave equation 263–4

wave equation 211
weighted residual method 268, 272